MW00723338

Radio / Tecn
Modifications
& Alignment Controls

WVRS3

Volume 6B

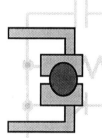

Dev.

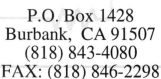

P.O. Box 1428
Burbank, CA 91507
(818) 843-4080
FAX: (818) 846-2298

Radio / Tech Modifications

NUMBER 6B   10 9 8 7 6 5 4 3 2 1

ISBN 0-917963-011-3 $19.95

Notice of Liability

artsci inc, books are available for bulk sales at quantity discounts. For information, please contact Marketing Manager, artsci inc, P.O. Box 1848, Burbank, CA 91507 FAX: (818) 846-2298

Printed in the United States of America

Contents

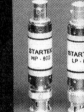

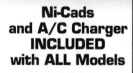
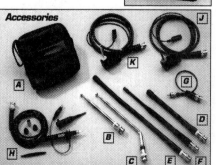

Preface

THERE ARE 2 BOOKS IN THIS VOLUME. AN ORDER FORM FOR THE OTHER BOOK IS AVAILABLE IN THE BACK OF THIS BOOK.

We call them Volume 6A and 6B. Volume 6A contains all known modifications for ICOM and Kenwood Radios and mods for the popular scanners. Volume 6B has all the modifications for Yeasu, Alinco, Standard, Azden, KDK, Ten Tec, Ranger, Uniden, Radio Shack and popular CB radios.

During the past 4 years we have created 6 volumes of Radio/Tech modifications. Each new volume included the information contained in the previous volumes. So if you have the current volume, you do not need to purchase the previous ones.

The illustrations are improved each year and the modifications have been performed by many people through out the world. The modifications contained in this book are accurate and current.

We make every effort to provide all available modifications for every radio we can find. We also try to keep the cost of the modification books as low as possible. We ask that you do not photocopy pages from these books. We will support you however we can, however, if you call us we will ask that you have the book in your hands at the time of the call.

It was only logical that we start to include the alignment points for each of the radios. Since you are inside them performing the modification, it is a good time to adjust the Modulation and Power levels. If you are not familiar with testing the levels, a section on service tools is provided to give you three methods of testing your radios.

If you find a new radio is not listed in these pages, contact us and ask about it. We may have a copy that did not make the printing deadline. If you purchased the book and have proof of purchase, we can make the new modification available to you.

Your comments and suggestions are always welcome. If the mod works great, let us know. If you can't make the mod work, let us know. We can't test every modification, we don't have all the radios. Your help will make the next volumes better for everyone.

TS-32P DIP Switch Programmable Encoder-Decoder $57.95

Universal design provides CTCSS capability to all FM transceivers. On-board DIP switch allows instant programming without tone elements, counters, or other test equipment. Crystal controlled for high accuracy and stability. The 32 location tone memory is complete with standard EIA tones from 67.0 to 203.5 Hz, or may be ordered with ANY 32 custom tone frequencies between 0 - 250.0 Hz (±0.1 Hz) at no extra charge. Multiple tone switching is easily done with your radio's channel select switch or separate single pole switch. A high pass tone rejection filter is included on board to remove tone from received audio. Reverse polarity protection and RF immunity are built in. Powered by 6 - 24 vdc, unregulated at 8ma. Supplied with color-coded wires terminated to plug directly onto the TS-32P. Mounting materials include hardware and double sided, insulated tape.

Programmable Encoder-Decoder
1.25"x 2.0"x 0.40"

TS-64 Microminiature CTCSS Encoder-Decoder $64.95

The latest - and smallest - programmable CTCSS encoder-decoder for use in FM transceivers. Ideal for many handheld radios and others with limited space. Select from 64 preset CTCSS tones between 33.0 Hz and 254.1 Hz using six PCB jumpers. Tone stability is crystal controlled with accuracy better than 0.05 Hz. Output level can be adjusted from OV to 3.0V. A time-out-timer feature permits programming transmit duration to eight different intervals decreasing "stuck mic" problems. Receiver Hi-pass filter and busy channel lockout are included. Decode sensitivity is 15mv. Power can be from 6.0vdc to 20.0vdc @ 9ma. Operating temperature range is from – 30°C to + 65°C. When P.T.T. switch is released, the TS-64 continues to key transmitter for 160ms. During this time, the TS-64 generates a reverse phase burst which will mute the decoding unit at the other end. A microminiature plug and socket with color coded wires attached is provided for hookup. Comes with double sided tape for quick mounting.

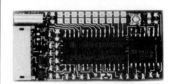

Microminiature CTCSS Encoder-Decoder
.78"x 1.70"x .25"

SS-32PA DIP Switch Programmable CTCSS Encoder $28.95

Universal design provides CTCSS encode capability to all FM transceivers. On-board DIP switch allows instant programming without tone elements, counters, or other test equipment. Crystal controlled for high accuracy and stability. The standard 32 tone memory contains the EIA tones from 67.0 to 203.5 Hz (or may be ordered with ANY 32 custom tone frequencies between 0 - 250.0 Hz at no extra charge). Multiple tone switching is easily achieved with your radio's channel select switch or a separate single pole switch.

SS-32PA / SS-32PB Encoder
0.9"x 1.3"x 0.4"

SS-32SMP Micro-Miniature CTCSS Encoder $27.95

Super small programmable CTCSS encoder for use in handheld radios and other size restricted applications. Has the same basic features as the SS-32PA (see above), but does not include the on-board DIP switch due to size limitations. Programming is done by soldering binary coded jumpers on the tone board.

SS-32SMP / SS-32SMP-B Encoder
0.53"x 1.00"x 0.16"

TE-64 Multi-purpose CTCSS/Burst Tone Encoder $79.95

Fully enclosed encoder provides, from a front dial rotary switch, all EIA CTCSS tones from 67.0 to 203.5 Hz PLUS all the common burst tones from 1600 to 2550 in 50 Hz increments. All available tone frequencies are permanently screened onto the faceplate, and selected with a calibrated dial. Great for test bench or service vehicle applications. Operates on 6-30 vdc, and all connections are made to a terminal strip at the rear of the unit. A 9 volt battery plug and cable is included, and may be attached at the terminal strip or soldered directly to the circuit board for field operation. Packaged in a high impact plastic case, with mounting bracket and hardware supplied.

TE-64 Tone Encoder
5.25"x 3.3"x 1.7"

TE-64D Multi-purpose CTCSS/Burst Tone Encoder w/Display $129.90

An enhanced version of the TE-64 encoder (see above). Features a two-digit LED which displays a number from 01 to 32 that in turn corresponds with the CTCSS or burst frequency selected by the front panel rotary switch. The two-digit number displayed is cross-referenced to the tone frequency on a chart located on the faceplate. Perfect for mobile applications, night-time operations, or whenever high visibility read-out is desired. Operates on 6-16 vdc (current draw does not allow operation from 9 volt battery).

TE-64D Tone Encoder w/Display
5.25"x 3.3"x 1.7"

NEW!

ID-8 Automatic Morse Station Identifier $89.95

Provides automatic Morse station identification for commercial, public safety, and amateur radio applications, including repeaters, base stations, mobiles, beacons, CW memory keyers, etc. Meets all FCC identification requirements. Low voltage/current operation and small size make it universally applicable. Low distortion, low impedance, adjustable sinewave output. High accuracy crystal control. All functions are programmable with plug-on keypad, included with each unit. Programmable options include: Eight selectable messages; CW speed 1-99 seconds; interval timer 1-99 minutes; hold off timer 0-99 seconds; CW tone frequency 100-3000 Hz; front porch delay interval 0-9.9 seconds; CW or MCW; etc. All programming is stored in a non-volatile EEPROM, which may be altered at any time via the included keypad. Supplied with programming keypad, wire set with microminiature plug for easy installation or removal, both hardware and tape mounting materials, and easy to follow instructions.

NEW!

ID-8 Automatic Morse Code Identifier
1.85"x 1.12"x 0.35"

COMMUNICATIONS SPECIALISTS, INC.
426 WEST TAFT AVENUE • ORANGE, CA 92665-4296
(714) 998-3021 • FAX (714) 974-3420
Entire U.S.A. (800) 854-0547 • FAX (800) 424-3420

Introduction

WHO SHOULD PERFORM MODIFICATIONS

This book is intended to be used as a reference guide for licensed Technicians. The text for each modification has been written with belief that the performing technician has experience with servicing modern radio equipment.

Attempts to perform these modifications by an inexperienced person may cause serious damage to the radio. Damage can occur by simply opening the radio case incorrectly. With the average repair cost of a damaged radio exceeding $150.00, it is a good investment paying a licensed technician to perform the modification.

Many of the new radios' components are barely larger than the head of a pin. Many of these parts require precision soldering. Excessive heat can damage these parts. Caution and the proper tools should be used to avoid damage to the components.

Some of the modifications presented in this book have not been tested. However, most of the modifications have been, at one time or another, reviewed by the technicians at the radio manufacturing or distributing plants.

USE THE PROPER EQUIPMENT

Alignment controls have been shown on many of the radios presented in the text. Proper alignment of a radio requires test equipment that is usually not available to the average operator. Exercise caution when changing the alignment controls. Improper settings can cause a radio to generate RF signals outside the desired frequency range. These undesired emissions will cause interference to others and may quite possibly be illegal.

Service manuals are valuable to any radio service technician. Service manuals will provide you with a list of components and detailed drawings of your radio. Our technical department is always looking to review the service manuals for the radios presented in this publication. If you have a service manual for a radio present here, we would like to review it.

MODIFICATIONS OF TYPE ACCEPTED EQUIPMENT

Some of the modifications presented in this publication may allow a radio to operate outside its design range. Using a radio outside its designed range may cause radio interference, equipment damage or may simply be illegal. Do not perform the modification if you have any concerns about the validity of the modification, or the purposes for a modified radio. Use your best judgment.

HOW FAR 'OUT OF BAND' WILL MY RADIO OPERATE?

The exact Receive and Transmit Frequency range of a radio are almost impossible to predict. The technicians at the factory tune a radio to operate in the specified range. Most radios can be tuned to operate almost anywhere within a 50 MHz range.

Once a radio is tuned, it should operate anywhere within a 30 MHz window. That's 15 MHz up and 15 MHz down from center. Most of the newer radios have been designed to allow a greatly increased range.

Your radio may operate better 'out of band' towards the bottom half of the workable range, and the next radio may operate better towards the top half.

The modifications presented here deal with opening up the microprocessors allowable frequency range. After a modification is complete, the microprocessor will tell the VCO/VXO circuitry what frequency to operate on. Can the current tuning of the RF coils and the circuitry operate at the desired frequency? That is the big question.

The tuning of the coils and VCO/VXO circuitry can be changed. These changes go well beyond the scope of this publication.

ACCURACY AND NEW MODIFICATIONS

The authors have made every attempt to present all the available modifications. As new radios and modifications become available, they will be added to the next publication. Outside contributions are accepted. A number of useful graphs, charts and tables are provided in the appendices.

Technicians are welcome to forward comments, suggestions and new modifications. Forward your modifications to our mailing address or FAX a copy to us.

SERVICE TOOLS

The cost of a service monitor, even the least expensive model, is over $2,000. You may be able to pick up a used unit for around $500. If you do manage to find a used service monitor, take it to be tested or calibrated on a new service monitor.

A service monitor performs a number of functions that are invaluable in aligning all types of radios. It can generate a signal on an exact frequency and allow you to control the signal strength and the amount of modulation applied to the signal. This feature will allow you to properly align the S-Meter and test the receiver sensitivity. A good receiver has a sensitivity of less than .2 micro volts.

Service Monitor

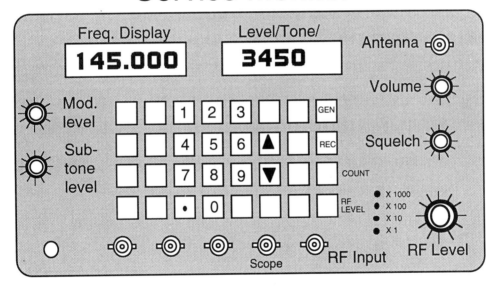

Perhaps the most valuable feature of a service monitor is its ability to act as a receiver and measure the frequency error & modulation.

Frequency error is measured in Hertz. A normal transmitter can be aligned or tuned up or down by as much as 5 kHz. (5,000 Hz). Most radios have an alignment control that will allow you to adjust the frequency up or down. A transmitter should be exactly on frequency. Within 200 Hz plus or minus is acceptable.

Measuring the modulation of a signal will allow you adjust the transmitter's microphone audio, DTMF pad and Sub-audible tone levels.

Suggested modulation levels:

Audio (microphone)	3,500 - 4,000 Hz
DTMF (touch tone)	3,000 - 3,500 Hz
PL (Sub-audible tone)	600-650 Hz

Alignment controls for these levels are available in most radios.

Other Valuable tools

There are a number of other tools that are a great deal less expensive than a service monitor. Most of these tools you should have in your tool box. If you do not have these tools, it a good idea that you invest a few dollars and pick them up.

Soldering iron

The modifications in this book require a 30-40 watt soldering iron. Make sure you have a small tip for the iron. A soldering gun is much too big. If you have one of the old guns, put it away until you are assembling a PL connector.

Some of the components used in the new radios are smaller than the letters in this sentence. You will need a steady hand and some experience desoldering components. A supply of solder braid is often the best method of removing a component.

Magnifying glass

Don't make a mistake here. The parts in the modern radios are small. You may not need one on some older radios, but open up one of the newer radios and you will wish one was handy.

Digital Volt/OHM Meter (DVM)

You must get one of these. They are handy for many things. Try to get one that has a continuity tone setting. An auto ranging meter is the best. If you can afford it, get one that has an auto shut off feature. Nothing is worse than grabbing your meter and finding the batteries are dead because you forgot to shut it off the last time you used it.

POOR MAN'S SERVICE TOOLS

If you are like most of us and can not afford a service monitor, there is a method available using inexpensive tools and a little help from a friend.

An oscilloscope is probably the most valuable instrument you can have. The cost of a new unit ranges from $250 up. A used unit can be purchased for as low as $50.

Scope

By connecting the receiver audio output (from the speaker jack) to the oscilloscope input, you can get an accurate visual view of the audio wave. With a little practice, you can accurately measure the audio levels.

If you are tuning up a transmitter, or the transmitter section of a transceiver, you will need the use of another receiver. If you have or can borrow a friend's handi-talkie, it will work just fine.

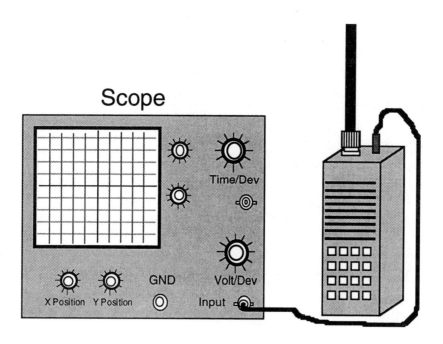

Scope

Connect your friends radio up to the oscilloscope through the external speaker jack. Turn the radio and oscilloscope on and adjust the receiver audio level to about 1/3. Turn the squelch off. Turn the Volt/Dev control to adjust the waves until they fill 1/2 of the display.

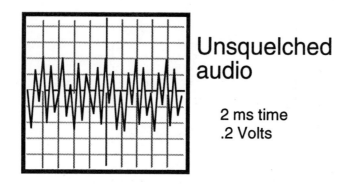

Unsquelched
audio

2 ms time
.2 Volts

There is a fine tuning control for both the Volt/Dev and Time/Div controls. Place them in the center position until you are ready to adjust the scope display discussed below.

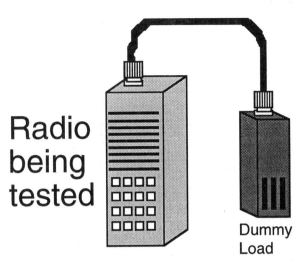

Now using your transmitter, press the PTT. (Make sure you are using a dummy load.) The scope display should appear below.

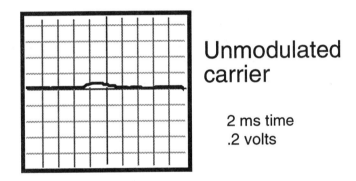

Unmodulated carrier

2 ms time
.2 volts

Now that you have the scope set up. Press the PTT key and talk into the microphone and watch the display. Hold the mic 3-4 inches away and say "FOUR" into the mic. Stretch the "FOUR" for 5 seconds.

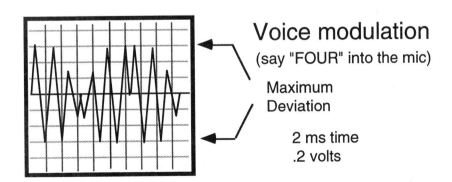

Voice modulation
(say "FOUR" into the mic)

Maximum
Deviation

2 ms time
.2 volts

The pattern on the scope is not as important as the maximum height of the wave crests.

The simplest testing method to see if your radio is accurately adjusted is to compare its signal to another radio that is operating properly. Transmit with the "GOOD" radio and adjust the scopes Volts/Div control to place the audio peaks at the markers as shown in the example above.

Now transmit with your radio and compare where the voice peaks are placed. If they are higher, adjust the Modulation/Deviation controls in your transmitter to a lower position. If they are lower, increase the control's position.

If possible, adjust the modulation/deviation control while you are transmitting and modulating.

You can adjust the levels of the DTMF key pad using the same method used on audio modulation. All DTMF tones have a rhythmic shape on the scope display. The DTMF tones will be lower in level that audio peaks. This is normal.

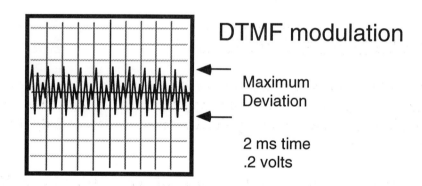

DTMF modulation

Maximum
Deviation

2 ms time
.2 volts

You can also adjust the level of the Sub-Audible PL tone using the scope. It will be necessary to adjust the Volt/row control to be more sensitive. A PL tone is only 20% the level of the voice modulation. Adjust the control to approx. 20 milli volts. Do not modulate the carrier with audio while you are adjusting the PL level.

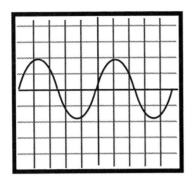

Sub-Audible carrier

2 ms time
20 m volts

Some receivers will filter out the sub-audible tone before it appears at the speaker jack. Most of the newer receivers do not do this so you should have no trouble watching the sub-audible wave form. If you can not get the expected wave form, check to make sure the transmitter is encoding PL tone. You should also check the receivers PL decode is turn off.

If you have gone this far, watch the display when you modulate a carrier that has a sub-audible tone. You will still see the tone no matter what type of modulation you use.

A more inexpensive method.

There is another method of checking the audio deviation levels using an audio VU meter. A VU meter can be purchased at your local Radio Shack. You can purchase the meter by itself, or in a case ready to hook up to your stereo.

Connect the VU meter to the speaker jack of your friend's radio or receiver.

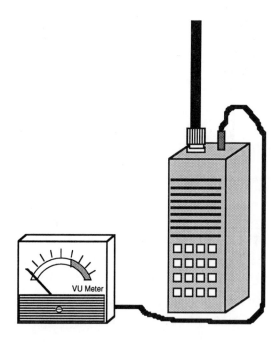

Using a properly working transmitter, transmit and hold down a DTMF tone key and adjust the receiver's volume control to cause the VU meter needle to set at the half-way point.

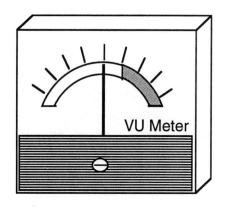

Adjust receiver volume to set meter at half-way position

Again press the PTT and measure where normal speaking audio causes the needle to peak.

Now using the radio to be tested, perform the same tests and adjust the transmitter's deviation controls to match the levels of the other radio.

This method will not work accurately enough to test sub-audible tone levels.

HEAR THE TONES...SEE THE NUMBERS!

*EIGHT DIGIT DISPLAY
*32 CHARACTER MEMORY
*ASCII SERIAL OUTPUT

MoTron TDD-8 TOUCH-TONE DECODER DISPLAY AND ASCII CONVERSION BOARD

The MoTron **TDD-8** is a wired and tested commercial touchtone test decoder board. The **TDD-8** decodes and displays all 16 touch-tone signals. The eight digit display, 32 character memory and left-right scroll function allows the user to capture and display up to 40 characters without loss of information. An ASCII serial output can be connected to a computer for automatic logging or remote data entry. The MoTron "Tonelog" IBM compatible software package is included with each TDD-8 at no additional charge. The computer interface cable can be purchased separately if needed.

Connect to almost any audio source - The MoTron **TDD-8** can be connected to a scanner, communications receiver, tape recorder, telephone answering machine etc.

IBM compatable software included - The **TDD-8** is a stand-alone device and does not need to be connected to a computer for decoding and displaying touch-tone digits. However, a serial ASCII output is provided on the board that can be connected to the RS-232 serial port of almost any computer. This allows you to use the **TDD-8** for numerous applications. The "Tonelog" IBM compatible sofware, that is included, will automatically log the date and time a number is decoded.

TDD-8 Touch-tone decoder with eight digit display, 32 character memory and ASCII serial output (wired/tested circuit board). **$109.00**

CAB-1 - Includes audio and computer cables. Audio patch-cord can be connected to most scanner and receiver speaker or earphone jacks. Mini phono plugs (3.5mm) on each end. Computer cable has mini phono (3.5mm) plug for connection to the TDD-8 and female DB-25 on the other end for computer connection. **$20.00**

PS-12 - 110VAC adapter. **$10.00**

PMK-1 - Plastic Mounting Kit. This is not a complete enclosure, but offers a simple means of protecting the board, making it easier to handle and use. Kit includes hard plastic sheets to cover the bottom and top of the board. Also included are rubber feet, spacers, nuts and bolts **$15.00**

ADD $5.00 FOR SHIPPING AND HANDLING. VISA/MC ACCEPTED.

Satisfaction guaranteed or your money back within 30 days of purchase (less shipping/handling). 90 day warranty on parts and labor.

Specifications: Board size: 6"X 2-3/8", Power requirements: 9 to 12 VDC @200 ma, DTMF response time: 40 ms (can decode fast auto-dialers), Audio input: 100 mv to 6 Vpp, Serial output: 1200 baud, 8 data bits, no parity.

MoTron Electronics
310 Garfield St., Suite 4
Eugene, OR 97402

ORDERS: 1-800-338-9058
INFO: (503) 687-2118
FAX: (503) 687-2492

(Touch-tone is a trademark of AT&T)

RIDING THE AIRWAVES WITH ALPHA & ZULU

Study for the Novice and No-Code Technician license tests with the newest comic book characters the Phoneticos.
112 comic strips review all the questions and answers.
288 pages, 8 1/2 X 11" format

U.S. REPEATER MAPBOOK #2

A repeater guide that shows where in each state principal open amateur repeaters are located. The Maps also show the important highways in each state. Tables showing the popular repeater in the states major cities are also presented. 2 meter, 200, 440 MHz and 1.2 GHz repeaters are shown. 144 pages, 6 x 9" format

FEDERAL ASSIGNMENTS Vol 3

The Frequency assignment master file.
The complete listing of all U.S. government used frequencies listed by agency and in frequency order.
Frequencies for Departments of: Agriculture, Air Force, Army, Commerce, Defence, Energy, Health and Human Services, Housing and Urban Developement, Interior, Justice, Labor, Navy, State, Treasury, Transportation and 29 Independent agencies & Commissions.
Over 350 pages, 8 1/2 X 11" format

AMATEUR HAMBOOK

Equipment & Log Sheets, Charts, Tables showing: worldwide callsigns, world times, shortwave listening frequencies, coax losses, CTCSS details, conversions, construction plans, emergency information, etc.
This book contains all the useful information a amateur radio operator needs to reference.
133 pages, 6 X 9" format.

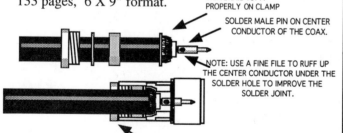

TRIM BRAID TO FIT PROPERLY ON CLAMP

SOLDER MALE PIN ON CENTER CONDUCTOR OF THE COAX.

NOTE: USE A FINE FILE TO RUFF UP THE CENTER CONDUCTOR UNDER THE SOLDER HOLE TO IMPROVE THE SOLDER JOINT.

PUSH PLUG BODY ON COAX AND TIGHTEN THE NUT.

QUICK-N-EASY SHORTWAVE LISTENING

What kind of radio to Buy? Whats a good antenna, What is there to listen to? This book contains pictures of radios, antenna construction and frequency lists. 6 X 9" format

art sci

P.O. Box 1428
Burbank, CA 91507
(818) 843-4080
FAX: (818) 846-2298

RADIO/TECH MODIFICATIONS # 5A

Modifications and alignment controls for ICOM & KENWOOD amateur radios, and UNIDEN, RADIO SHACK & REGENCY Scanners. Over 200 pages, 8 1/2" X 11"

RADIO/TECH MODIFICATIONS # 5B

Modifications and alignment controls for ALINCO, YAESU, STANDARD, AZDEN radios and 10 meter & CB radios.

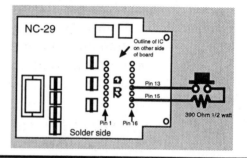

HAM RADIO RESOURCE GUIDE

For Southern California only. A booklet of all the information an amateur radio operator needs.
Listings of clubs, testing centers, sario stores and surplus dealers. Maps of repeaters, store and swap meet locations. Listing of Packet repeaters, phone BBS and node lists. 64 pages 8 1/2" X 5 1/2"

NORTH AMERICAN SHORTWAVE FREQUENCY GUIDE

Accurate and complete listing of all English and Spanish broadcasts on the 0 - 30 MHz shortwave bands. Listing are presented in frequency order.
Over 200 pages, 8 1/2" X 11"

LOST USER MANUALS

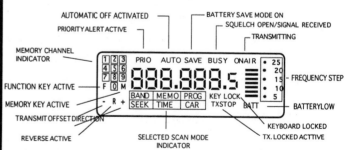

Lost the manual for your HT or Mobile rig? Did you purchase a used radio and it did not come with a manual? Do you have the manual but still can work the radio quickly?
"LOST USERS MANUALS" contains operating instructions for all the popular amateur radios and scanners. ICOM, Yaesu, Kenwood, Alinco, Standard, Uniden and other manufactors radios. Each radio is given 2 to 5 pages of drawing, charts and programming instructions. Over 140 Pages, 8 1/2 X 11" format.

Radio / Tech Modifications

ALINCO Radio Modifications

ALINCO ALD-24T

EXPANDED RF

1. Remove Battery and Antenna.
2. Remove top and bottom covers.
3. Remove Main dial,Vol & SQL knobs. Remove the retaining rings.
4. Remove front cover to access front panel circuit board.
5. **Solder bridge four sets of pads as shown**.
6. Reassemble radio.
7. Reset microprocessor (Press reset button)

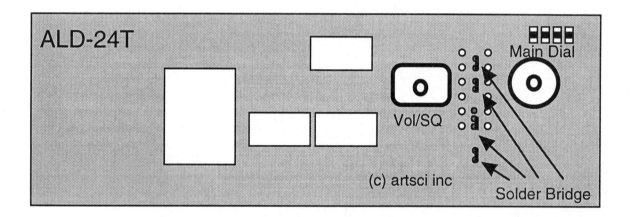

MORE ---

ALINCO ALD-24T

ALIGNMENT CONTROLS

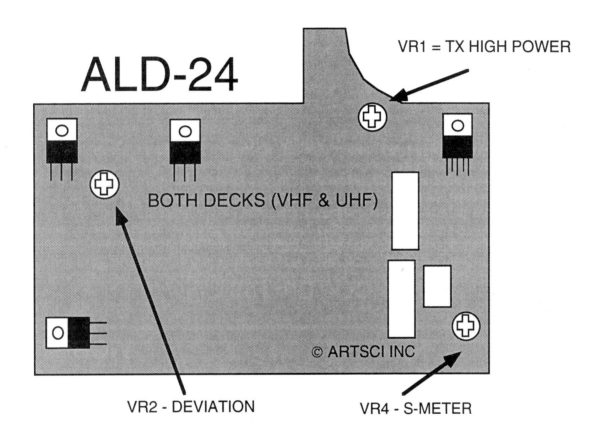

VR1 = TX HIGH POWER

ALD-24

BOTH DECKS (VHF & UHF)

© ARTSCI INC

VR2 - DEVIATION

VR4 - S-METER

ALINCO ALR-22T

EXPANDED RF

1. Remove Battery and Antenna.
2. Remove top and bottom covers.
3. Remove Main dial,Vol & SQL knobs. Remove the retaining rings.
4. Remove front cover to access front panel circuit board.
5. **Solder bridge Three sets of pads as shown.**
6. Reassemble radio

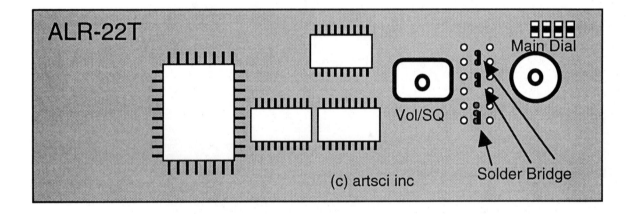

MORE ---

ALINCO ALR-22T

ALIGNMENT CONTROLS

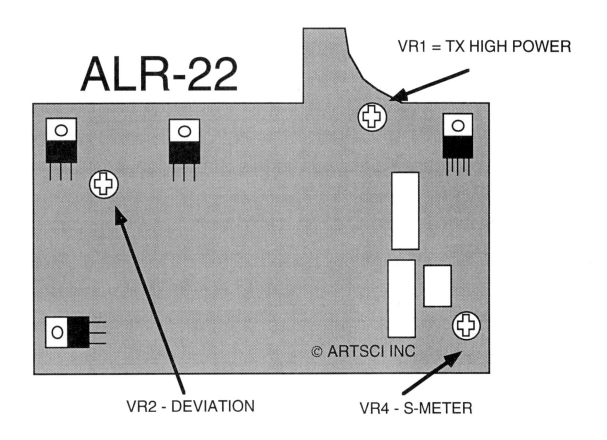

ALR-22

VR1 = TX HIGH POWER

© ARTSCI INC

VR2 - DEVIATION

VR4 - S-METER

MORE ---

Performance Report

Radio _____   Date _____

Owner : Name _____
        Address _____
        City _____ St. ____ Zip _____
        Phone (____) ____ - _____

Description	Before	After
Power out (Low)	_____ Watts	_____ Watts
Power out (High)	_____ Watts	_____ Watts
Frequency Error (Simplex)	_____ Hz	_____ Hz
Frequency Error (Offset)	_____ Hz	_____ Hz
Receive Sensitivity (Mid-band)	_____ uv	_____ uv
Receive Sensitivity (____MHz)	_____ uv	_____ uv
Receive Sensitivity (____MHz)	_____ uv	_____ uv
PL Deviation	_____ Hz	_____ Hz
DTMF Deviation	_____ KHz	_____ KHz
Audio Deviation	_____ KHz	_____ KHz
Lowest usable Freq @ .5 Pwr	_____ MHz	_____ MHz
Highest usable Freq @ .5 Pwr	_____ MHz	_____ MHz

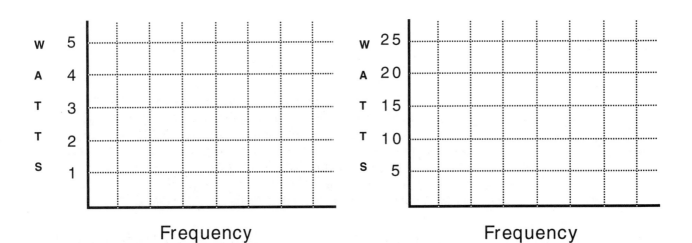

Frequency   Frequency

ALINCO ALR-22T

MICROPHONE MOD

1. Remove Battery and Antenna.
2. Remove top and bottom covers.
3. Remove Main dial,Vol & SQL knobs. Remove the retaining rings.
4. Remove front cover to access front panel circuit board.
5. Locate and remove the Microphone Green, Orange & Purple wires.
6. **Solder the wired as shown in drawing**
7. Reassemble radio.
8. Open Microphone.
9. Remove the Ground side of the Up/Down buttons and tie them together.
10. Connect the Orange wire to the two tied wires.
11. Reassemble Microphone.

ALR-22T
Mic/Memory
UP/Down
Mod

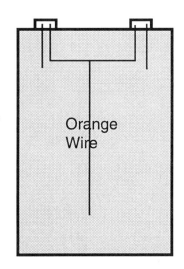

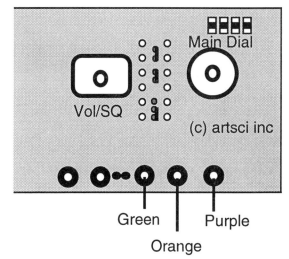

ALINCO DJ-100T

EXPANDED RF

1. Remove Battery and Antenna.
2. Remove screws from case and open radio.
3. Locate & **Cut Jumpers per drawing.**
4. **Clip pin 2 on IC401(S7116A) and connect it to pin 14** (for simplex PL tone)
   This chip is located on the TONE SW board.
4. Reassemble radio.
5. Reset Micro Processor.

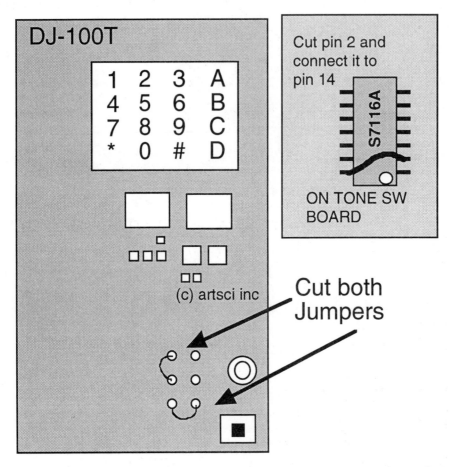

DJ-100T

1	2	3	A
4	5	6	B
7	8	9	C
*	0	#	D

(c) artsci inc

Cut pin 2 and connect it to pin 14

S7116A

ON TONE SW BOARD

Cut both Jumpers

MORE ---

ALINCO DJ-100T
ALIGNMENT CONTROLS

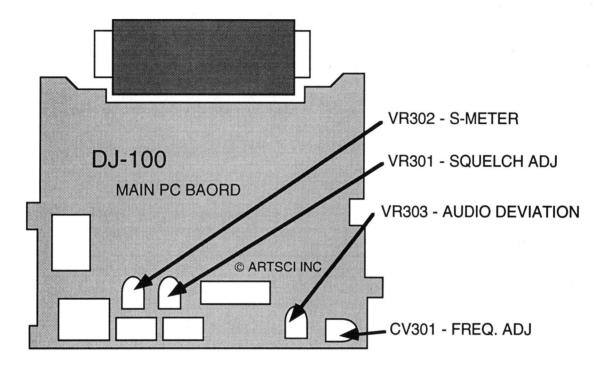

VR302 - S-METER

VR301 - SQUELCH ADJ

VR303 - AUDIO DEVIATION

CV301 - FREQ. ADJ

DTMF DEV (VR601) ON DTMF BOARD

VR401- PL DEVIATION

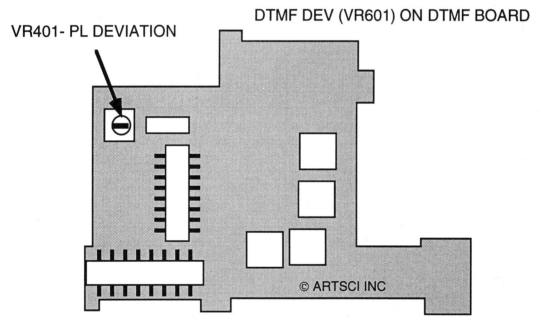

ALINCO DJ-120T

EXPANDED RF

1. Remove Battery and Antenna.
2. Remove screws from case and open radio.
3. Locate & Cut Jumpers per drawing.
4. **Clip pin 2 on IC401(S7116A) and connect it to pin 14** (for simplex PL tone)
   This chip is located on the TONE SW board.
4. Reassemble the radio.
5. Reset the microprocessor.

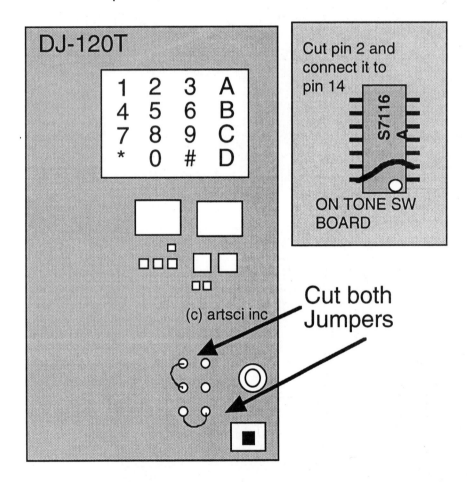

MORE ---

ALINCO DJ-120T
ALIGNMENT CONTROLS

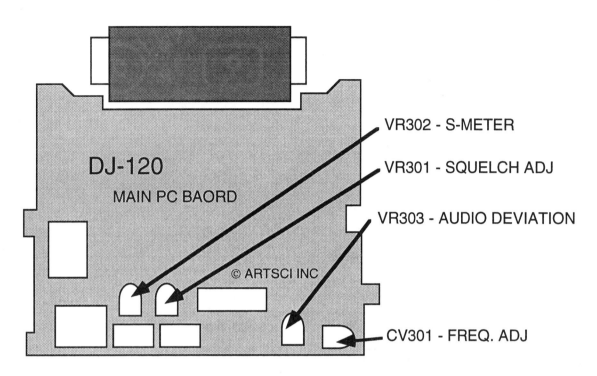

DJ-120

MAIN PC BAORD

© ARTSCI INC

VR302 - S-METER

VR301 - SQUELCH ADJ

VR303 - AUDIO DEVIATION

CV301 - FREQ. ADJ

DTMF DEV (VR601) ON DTMF BOARD

VR401- PL DEVIATION

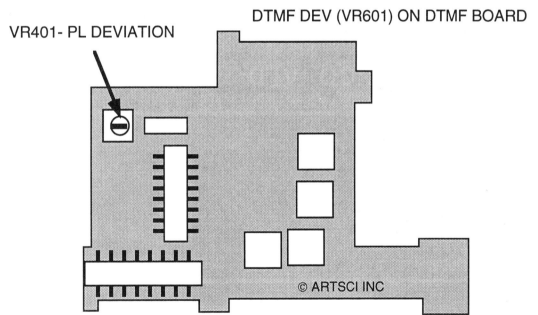

© ARTSCI INC

ALINCO DJ-160T

EXPANDED RF

1. Remove Battery and Antenna.
2. Remove 2 screws back of case and four screws from battery slide clip.
3. Remove Main dial,Vol & SQL knobs. Remove the retaining rings.
4. Remove the top cover.
5. Open radio.
6. Locate and **cut yellow wire** behind the battery release button.
7. Reassemble radio.
8. **Reset microprocessor.** (Press and hold [F] key and turn power on.)

DJ-160T

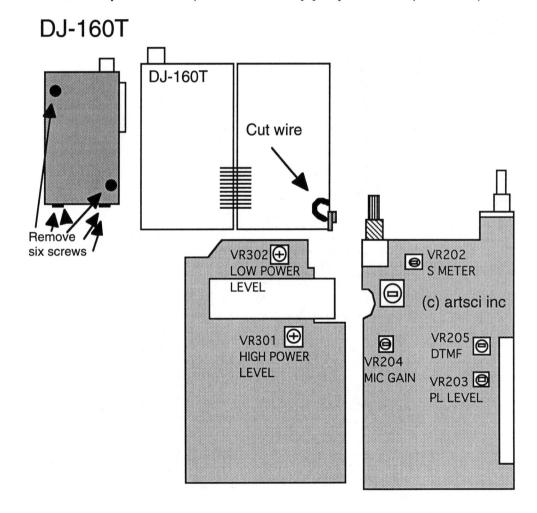

ALINCO DJ-162T

EXPANDED RF

1. Remove Battery and Antenna.
2. Remove 2 screws back of case and four screws from battery slide clip.
3. Remove Main dial,Vol & SQL knobs. Remove the retaining rings.
4. Remove the top cover.
5. Open radio.
6. Locate and **cut yellow wire** behind the battery release button.
7. Reassemble radio.
8. **Reset microprocessor.** (Press and hold [F] key and turn power on.)

DJ-162T

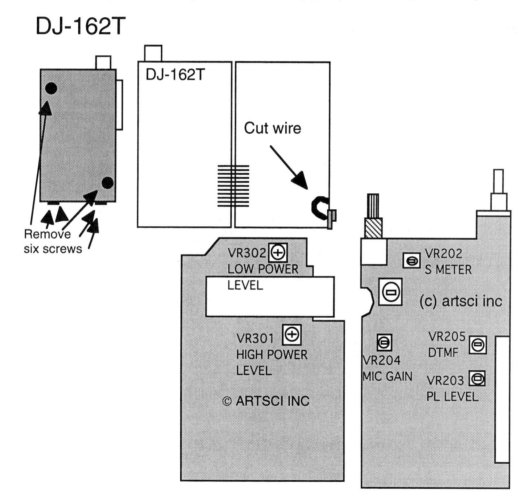

ALINCO DJ-180T

EXPANDED RF / ALIGNMENTR CONTROLS

This mod will void the warrenty.

1. Remove Battery and Antenna.
2. Remove the four screws holding the battery slide plate in location.
   (Careful not to break the battery plate wires)
3. Locate and **cut the "PINK" wire**. (Only the PINK wire)
4. Reassemble the unit.
5. **Reset the microprocessor**
   (Press and hold the [LAMP] button and turn the power on.)

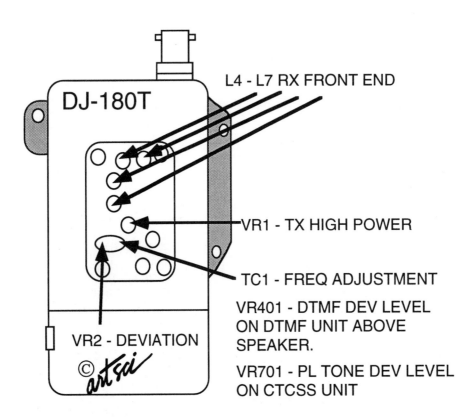

L4 - L7 RX FRONT END

VR1 - TX HIGH POWER

TC1 - FREQ ADJUSTMENT

VR401 - DTMF DEV LEVEL
ON DTMF UNIT ABOVE
SPEAKER.

VR701 - PL TONE DEV LEVEL
ON CTCSS UNIT

VR2 - DEVIATION

Optional Receive only mod: (130 - 173 MHz)

1. Reset the microprocessor
   (Press and hold the [LAMP] button and turn the power on.)

ALINCO DJ-460T

EXPANDED RF

1. Remove Battery and Antenna.
2. Remove 2 screws back of case and four screws from battery slide clip.
3. Remove Main dial,Vol & SQL knobs. Remove the retaining rings.
4. Remove the top cover.
5. Open radio.
6. Locate and **cut wire behind the battery release button.**
7. Reassemble radio.
8. **Reset microprocessor.** (Press and hold [F] key and turn power on.

DJ-460

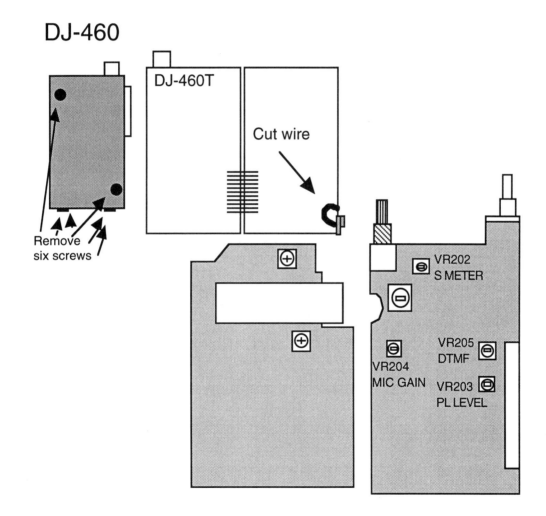

ALINCO DJ-500T

EXPANDED RF

1. Remove Battery and Antenna.
2. Remove screws from case (3 Long & 2 short)
3. **Remove green component per drawing.**
4. Reassemble radio.
5. Reset the radio. (Reset switch is located below the PTT Switch

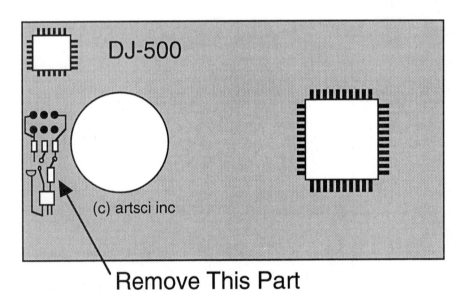

Remove This Part

MORE ---

ALINCO DJ-500T

ALIGNMENT CONTROLS

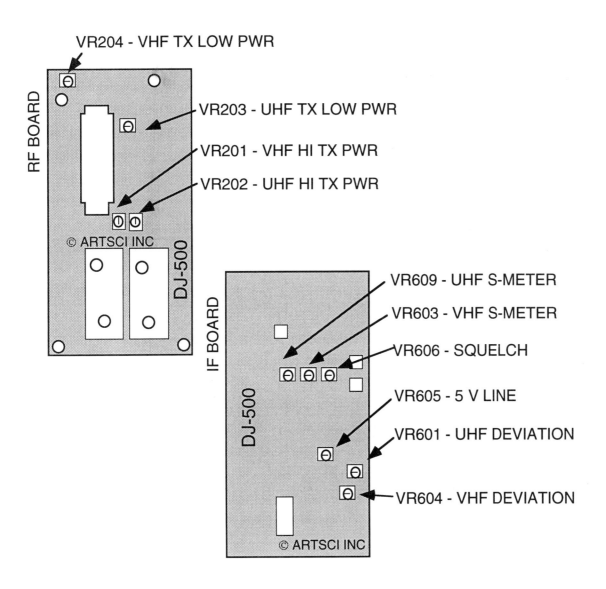

VR204 - VHF TX LOW PWR

VR203 - UHF TX LOW PWR

VR201 - VHF HI TX PWR

VR202 - UHF HI TX PWR

VR609 - UHF S-METER

VR603 - VHF S-METER

VR606 - SQUELCH

VR605 - 5 V LINE

VR601 - UHF DEVIATION

VR604 - VHF DEVIATION

RF BOARD

IF BOARD

© ARTSCI INC

DJ-500

VR1 - DTMF DEVIATION (ON CPU BOARD)

ALINCO DJ-560

EXPANDED RF

1. Remove battery and antenna.
2. Remove screws from back of case.
3. Remove all 4 screws from battery plate.
4. Remove screw next to the BNC connector.
5. Remove the Dial, UHF and VHF knobs.
6. Unscrew the Lock rings under the Dial, UHF and VHF knobs.
7. Remove the top cover.
8. Remove the 4 screws hold in the radio together.
4. Locate and **cut orange wire** directly below the PTT switch.
    (Some units have a yellow wire)
5. Reassemble the radio.
6. **Reset the CPU.** (Press and hold [FUNCTION] and turn power on)

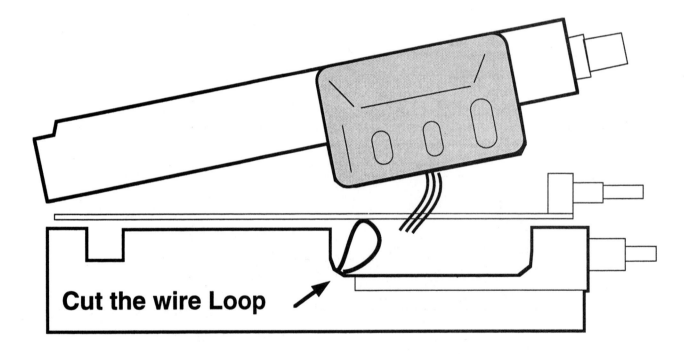

Cut the wire Loop

MORE ---

ALINCO DJ-560

ALIGNMENT CONTROLS

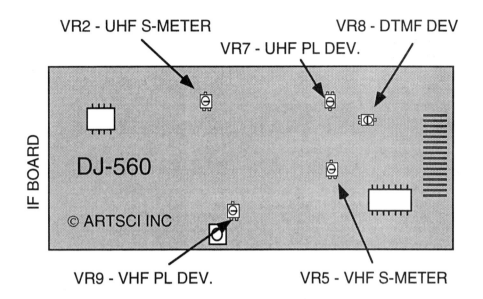

VR2 - UHF S-METER

VR7 - UHF PL DEV.

VR8 - DTMF DEV

IF BOARD

DJ-560

© ARTSCI INC

VR9 - VHF PL DEV.

VR5 - VHF S-METER

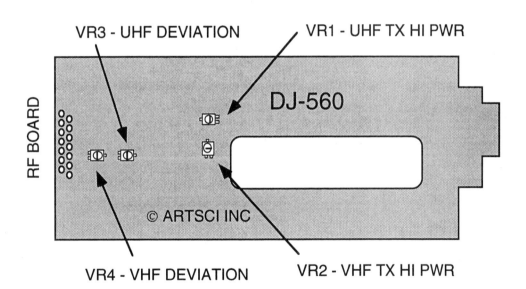

VR3 - UHF DEVIATION

VR1 - UHF TX HI PWR

RF BOARD

DJ-560

© ARTSCI INC

VR4 - VHF DEVIATION

VR2 - VHF TX HI PWR

ALINCO DJ-580T

EXPANDED RF
Aircraft Band RX & 800 MHZ RX

1. Remove battery and antenna.
2. Remove the four (4) screws on the bottom of the radio.
3. Remove the battery slide plate.
4. Locate and **CUT the BLUE wire** (for expanded RF)
5. Locate and **CUT the RED wire** (for aircraft and 800 MHZ RX.
6. Reassemble the radio.
7. **Reset the microprocessor.**
   (Press and hold the [FUNCTION] key and turn the radio on).

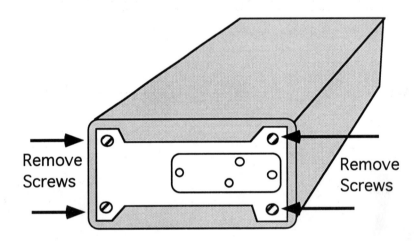

Remove Screws

Remove Screws

To Select the AIRCRAFT BAND
Press the [FUNCTION] and [VHF] key simultaneously.
The Letter "A" (AM mode) will appear on the display.
(press again to select the 2 meter band)

To Select the 800 MHz BAND
Press the [FUNCTION] and [UHF] key simultaneously.
(press again to select the 440 MHz band)

MORE ---

ALINCO DJ-580T

ALIGNMENT CONTROLS

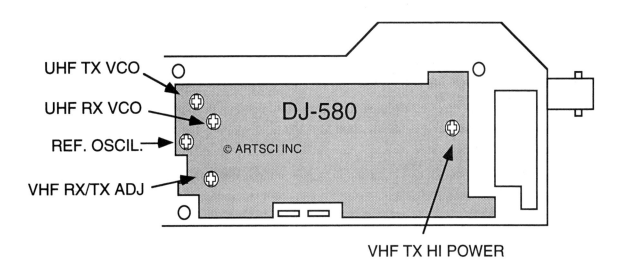

UHF TX VCO
UHF RX VCO
REF. OSCIL.
VHF RX/TX ADJ

DJ-580
© ARTSCI INC

VHF TX HI POWER

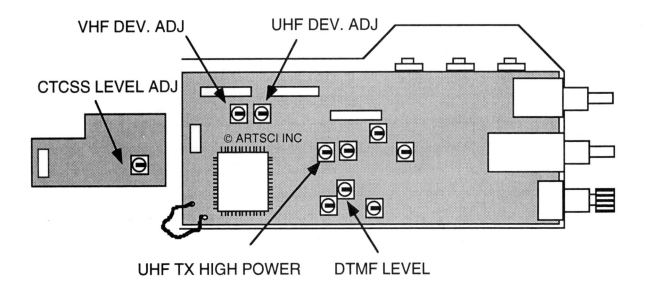

VHF DEV. ADJ   UHF DEV. ADJ

CTCSS LEVEL ADJ

© ARTSCI INC

UHF TX HIGH POWER   DTMF LEVEL

ALINCO DJ-F1T

EXPANDED RF

1. Remove battery and antenna.
2. Remove 5 screws from the back of the case.
3. Slide and hold the Battery lock button open the radio carefully.
4. Locate and **cut the RED jumper wire**. (AM airband reception)
5. Locate and **cut the BLUE jumper**. (Expanded RF)
6. Reassemble the radio.
7. **Reset the microprocessor.** (Press and hold the [F] key and turn the power on)

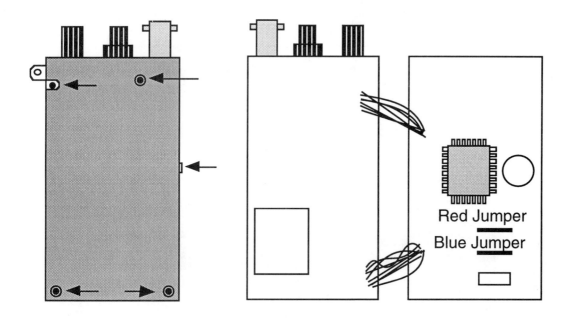

Red Jumper

Blue Jumper

TURN ON/OFF AIRBAND: Press the [B] key. an "A" will appear on the display to indicate the AM mode is operating.

MORE ---

ALINCO DJ-F1T

ALIGNMENT CONTROLS

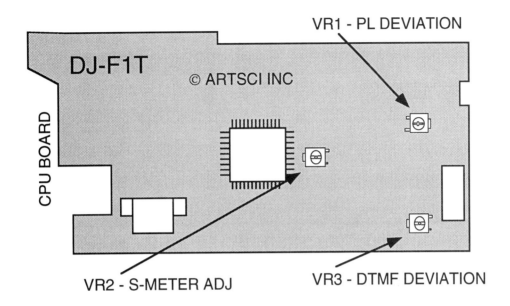

VR1 - PL DEVIATION

DJ-F1T

© ARTSCI INC

CPU BOARD

VR2 - S-METER ADJ

VR3 - DTMF DEVIATION

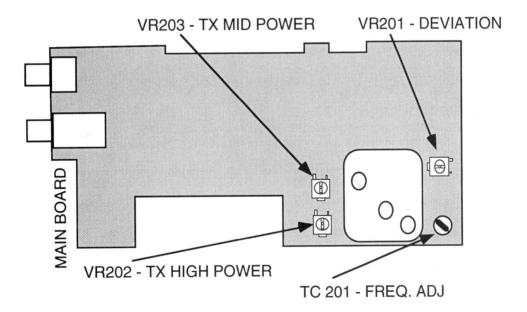

VR203 - TX MID POWER

VR201 - DEVIATION

MAIN BOARD

VR202 - TX HIGH POWER

TC 201 - FREQ. ADJ

ALINCO DR-110T

EXPANDED RF & ALIGNMENT CONTROLS

1. Remove Power and Antenna.
2. Remove screws from top case and open radio.
3. **Cut the yellow wire** on the control board.
4. Reassemble radio
5. **Reset microprocessor.** (Turn radio on. Press and hold [F] and [VFO/M] and turn power off and while still holding keys, turn power back on.

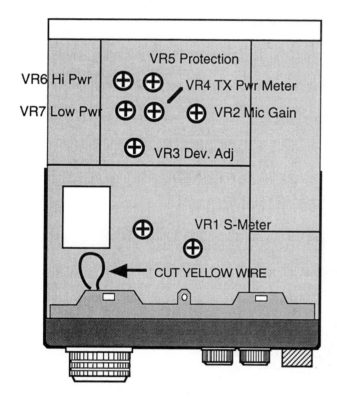

VR6 Hi Pwr
VR7 Low Pwr
VR5 Protection
VR4 TX Pwr Meter
VR2 Mic Gain
VR3 Dev. Adj
VR1 S-Meter
CUT YELLOW WIRE

VR1 on
Tone squelch
board the
PL Level

ALINCO DR-112T

EXPANDED RF & ALIGNMENT CONTROLS

1. Remove Power and Antenna.
2. Remove screws from top case and open radio.
3. **Cut the yellow wire** on the control board
4. Reassemble radio
5. **Reset microprocessor.** (Turn radio on. Press and hold [F] and [VFO/M] and turn power off and while still holding keys, turn power back on.

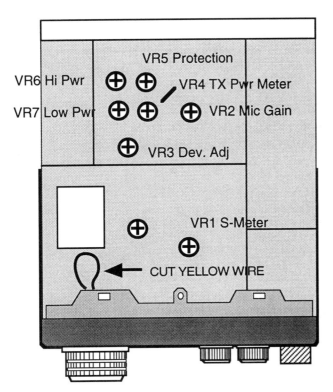

VR1 on
Tone squelch
board the
PL Level

ALINCO DR-119T

ALIGNMENT CONTROLS

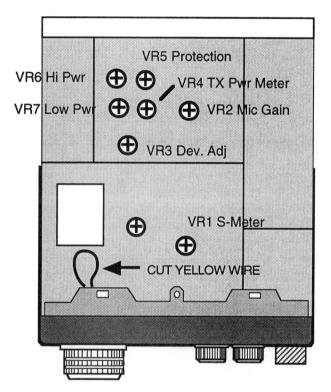

VR5 Protection

VR6 Hi Pwr

VR4 TX Pwr Meter

VR7 Low Pwr

VR2 Mic Gain

VR3 Dev. Adj

VR1 on
Tone squelch
board the
PL Level

VR1 S-Meter

CUT YELLOW WIRE

ALINCO DR-130T
EXPANDED RF

1. Remove power and Antenna.
2. Remove the top cover.
3. Locate and cut the BLUE jumper wire.
4. Reassemble the radio
5. Reset the microprocessor.
      (Press and hold the [FUNCTION] button and turn the radio on)

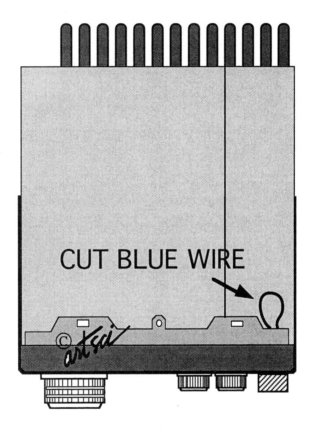

CUT BLUE WIRE

ALINCO DR-510T

EXPANDED RF / CROSS BAND REPEATER MOD

1.   Remove Battery and Antenna.
2.   Remove screws from case and open radio.
3.   Cut the yellow wire looped around the blue condenser
4.   Remove 2 screws from corners of tone board, to expose motherboard.
5.   **Solder a 16 volt 100uf electrolytic** as shown. (note 10-100uf)
     - lead to pin 8 of M54959P   + lead to third pin of socket (Orange wire)
6.   Remove the front cover
7.   **Short chip resistor R35 and solder bridge the pads** to the left of the resistor.
8.   Reassemble radio
9.   Reset microprocessor (Push reset button)

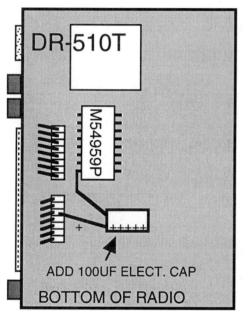

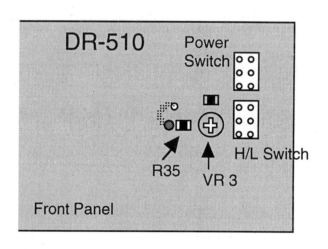

CROSS BAND REPEATER PROCEEDURES - Select the VHF & UHF
frequencies and press [SHIFT] until "DUAL" appears.

TURN ON : Press and hold [REV] and turn power on. The volume control controls the amount of repeater audio.

TURN OFF : Turn radio off.

MORE ---

ALINCO DR-510T

ALIGNMENT CONTROLS

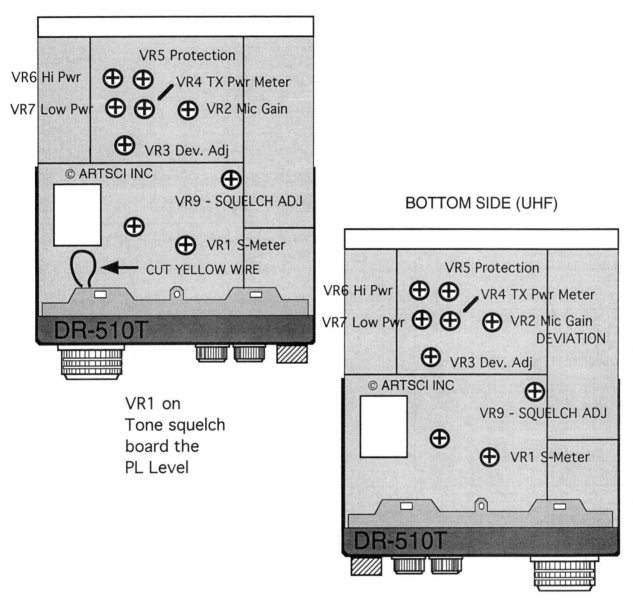

UPPER SIDE (VHF)

VR5 Protection

VR6 Hi Pwr ⊕ ⊕ VR4 TX Pwr Meter

VR7 Low Pwr ⊕ ⊕ ⊕ VR2 Mic Gain

⊕ VR3 Dev. Adj

© ARTSCI INC ⊕

VR9 - SQUELCH ADJ

⊕

⊕ VR1 S-Meter

← CUT YELLOW WIRE

DR-510T

VR1 on
Tone squelch
board the
PL Level

BOTTOM SIDE (UHF)

VR5 Protection

VR6 Hi Pwr ⊕ ⊕ VR4 TX Pwr Meter

VR7 Low Pwr ⊕ ⊕ ⊕ VR2 Mic Gain
DEVIATION

⊕ VR3 Dev. Adj

© ARTSCI INC ⊕

VR9 - SQUELCH ADJ

⊕

⊕ VR1 S-Meter

DR-510T

ALINCO DR-570T

EXPANDED RF / CROSS BAND REPEATER MOD

1. Remove Battery and Antenna.
2. Remove screws from case and open radio (3 screws in the top and 2 in the bottom.)
3. Locate and cut the indicated component. see drawing
4. **Turn repeater/normal switch to repeater mode.**
5. **Reset the microprocessor.** (Press and hold [FUNCTION] and turn power on)
6. Remove the two pin connector to disable audio bleed in repeater mode.
7. Reassemble radio.

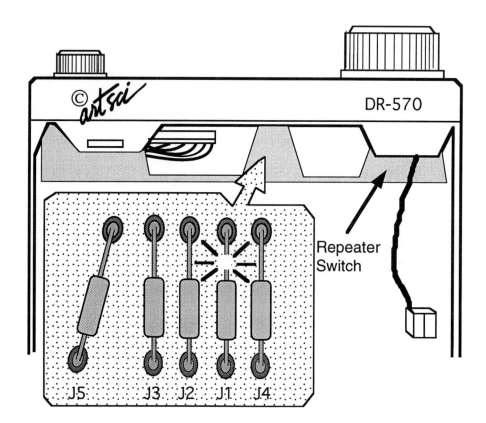

MORE ---

ALINCO DR-570T

ALIGNMENT CONTROLS

ALIGNMENT	UHF	VHF
TX HIGH POWER	VR5	VR2
TX LOW POWER	VR7	VR4
RF METER	VR6	VR1
DEVIATION	VR3	VR3
SQUELCH ADJ	VR1	VR1
S-METER	VR2	

Performance Report

Radio _____   Date _____

Owner : Name _____
         Address _____
         City _____ St. _____ Zip _____
         Phone (_____) _____ - _____

Description	Before	After
Power out (Low)	_____ Watts	_____ Watts
Power out (High)	_____ Watts	_____ Watts
Frequency Error (Simplex)	_____ Hz	_____ Hz
Frequency Error (Offset)	_____ Hz	_____ Hz
Receive Sensitivity (Mid-band)	_____ uv	_____ uv
Receive Sensitivity (____MHz)	_____ uv	_____ uv
Receive Sensitivity (____MHz)	_____ uv	_____ uv
PL Deviation	_____ Hz	_____ Hz
DTMF Deviation	_____ KHz	_____ KHz
Audio Deviation	_____ KHz	_____ KHz
Lowest usable Freq @ .5 Pwr	_____ MHz	_____ MHz
Highest usable Freq @ .5 Pwr	_____ MHz	_____ MHz

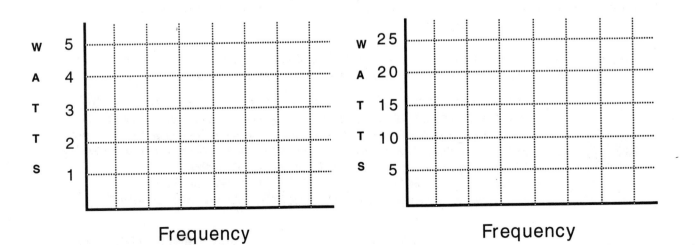

Frequency   Frequency

ALINCO DR-590T
EXPANDED RF
CROSS BAND REPEATER MOD

1. Remove Power and Antenna.
2. Remove the four screws, (2 on each side) holding the LCD display to the main body of the radio.
3. DO NOT DISCONNECT THE BLACK CONNECTOR CABLE FROM THE LCD DISPLAY.
4. Locate and unscrew the 2 screws holding the LCD display together.
5. Carefully separate the back cover of the display from the front cover. Use a flat blade screwdriver to apply slight pressure to the locking tabs in the top of the display.
6. Locate and **cut the loop of BLUE wire.**
7. Reassemble the radio.
8. **Reset the microprocessor.** (Press and hold the [FUNCTION] key and turn power on.)

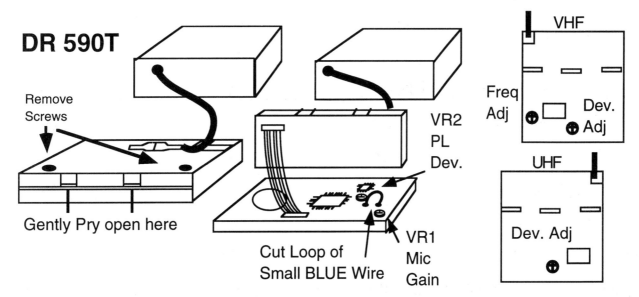

DR 590T

Remove Screws

Gently Pry open here

Cut Loop of Small BLUE Wire

VR1 Mic Gain

VR2 PL Dev.

VHF — Freq Adj — Dev. Adj

UHF — Dev. Adj

ENABLE REPEATER MODE: Simultaneously press the [FUNCTION] key and the [VHF] Key. The display will alternate between VHF and UHF every 3 seconds.

DISABLE REPEATER MODE: Simultaneously press the [FUNCTION] key and the [UHF] Key.

A audio frequency response kit is available from Alinco. Contact them for the parts and instruction sheet. (This is for improving the Cross-band repeater audio)

ALINCO DR-599T
EXPANDED RF / CROSS BAND REPEATER MOD

1.   Remove the Power cable and Antenna.
2.   Remove the 4 screws, (two on each side).
       HOLD THE CONTROL HEAD against the main unit.
3.   Remove the 2 screws holding the control head together.
4.   Carefully seperate the back and front cover of the control unit.
5.   **Cut the RED wire** to allow reception in the Aircraft and the 800 MHz band.
6.   **Cut the BLUE wire** to expand the TX & RX frequencies.
7.   Reasseble the control head.
8.   Remove the bottom cover. (two additional screws on the bottom cover)
9.   For 800 MHz RX, feed a new antenna cable thru the optional antenna jack on the
       back of the main body of the radio.
10.   Locate antenna connector CN59 and attach the antenna cable.
11.   Reassemble the radio .
12.   **Reset the Microprocessor.** (Push and hold the [FUNC] key and turn the power on)

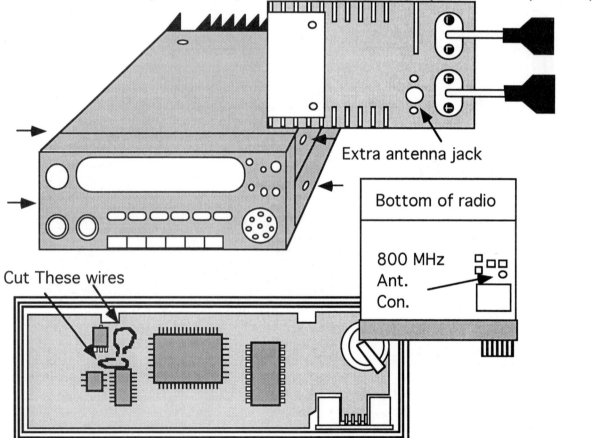

Extra antenna jack

Bottom of radio

800 MHz
Ant.
Con.

Cut These wires

ALINCO DR-599T
ALIGNMENT CONTROLS

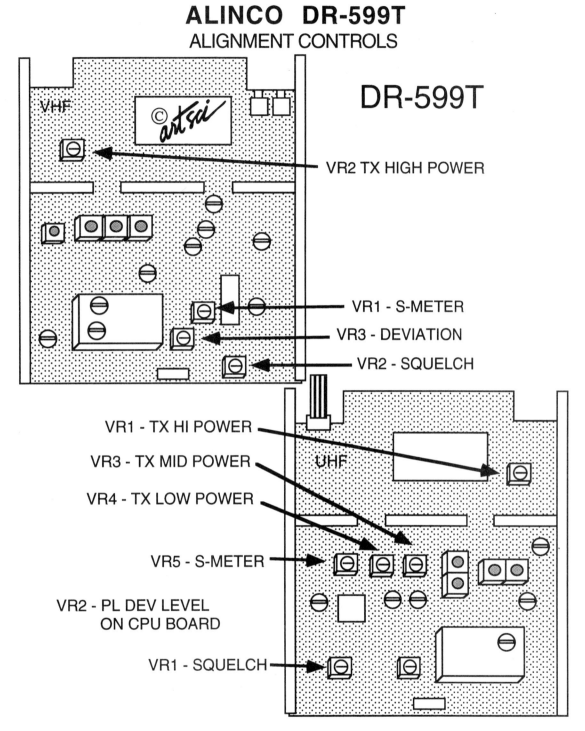

DR-599T

VR2 TX HIGH POWER

VR1 - S-METER

VR3 - DEVIATION

VR2 - SQUELCH

VR1 - TX HI POWER

VR3 - TX MID POWER

VR4 - TX LOW POWER

VR5 - S-METER

VR2 - PL DEV LEVEL
ON CPU BOARD

VR1 - SQUELCH

ALINCO DR-600T
EXPANDED RF / CROSS BAND REPEATER MOD

1. Remove the Power cable and Antenna.
2. Remove the 4 screws, (two on each side).
      HOLD THE CONTROL HEAD against the main unit.
3. Remove the 2 screws holding the control head together.
4. Carefully seperate the back and front cover of the control unit.
5. **Cut the RED wire** to allow reception in the Aircraft and the 800 MHz band.
6. **Cut the BLUE wire** to expand the TX & RX frequencies.
7. Reasseble the control head.
8. Remove the bottom cover. (two additional screws on the bottom cover)
9. For 800 MHz RX, feed a new antenna cable thru the optional antenna jack on the
      back of the main body of the radio.
10. Locate antenna connector CN59 and attach the antenna cable.
11. Reassemble the radio .
12. **Reset the Microprocessor**. (Push and hold the [FUNC] key and turn the power on)

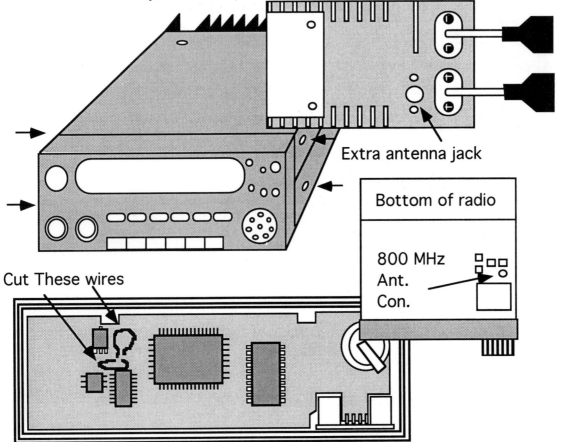

Extra antenna jack

Bottom of radio

800 MHz
Ant.
Con.

Cut These wires

ALINCO DR-1200T

ALIGNMENT CONTROLS

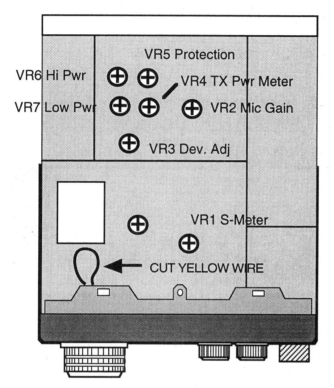

VR6 Hi Pwr

VR5 Protection

VR4 TX Pwr Meter

VR7 Low Pwr

VR2 Mic Gain

VR3 Dev. Adj

VR1 S-Meter

CUT YELLOW WIRE

VR1 on
Tone squelch
board the
PL Level

ALINCO HT

Packet Connections

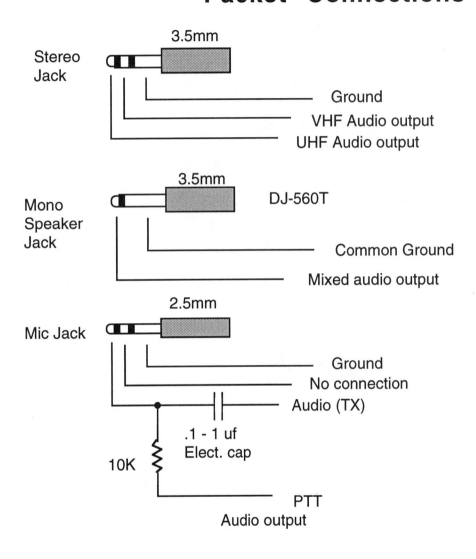

Stereo Jack — 3.5mm
- Ground
- VHF Audio output
- UHF Audio output

Mono Speaker Jack — 3.5mm — DJ-560T
- Common Ground
- Mixed audio output

Mic Jack — 2.5mm
- Ground
- No connection
- Audio (TX)
- .1 - 1 uf Elect. cap
- 10K
- PTT
- Audio output

ALINCO Mobile

Packet Connections

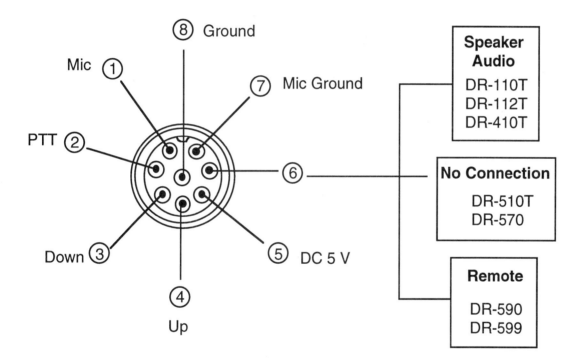

Performance Report

Radio _____ Date _____

Owner : Name _____
        Address _____
        City _____ St. _____ Zip _____
        Phone (____) _____ - _____

Description	Before	After
Power out (Low)	_____ Watts	_____ Watts
Power out (High)	_____ Watts	_____ Watts
Frequency Error (Simplex)	_____ Hz	_____ Hz
Frequency Error (Offset)	_____ Hz	_____ Hz
Receive Sensitivity (Mid-band)	_____ uv	_____ uv
Receive Sensitivity (____MHz)	_____ uv	_____ uv
Receive Sensitivity (____MHz)	_____ uv	_____ uv
PL Deviation	_____ Hz	_____ Hz
DTMF Deviation	_____ KHz	_____ KHz
Audio Deviation	_____ KHz	_____ KHz
Lowest usable Freq @ .5 Pwr	_____ MHz	_____ MHz
Highest usable Freq @ .5 Pwr	_____ MHz	_____ MHz

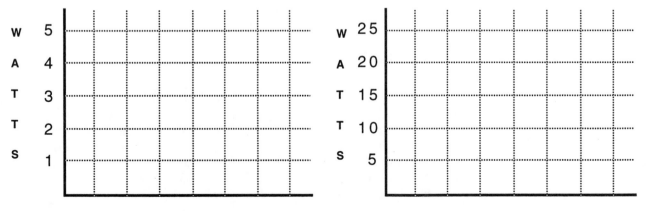

W A T T S 5 4 3 2 1

Frequency

W A T T S 25 20 15 10 5

Frequency

Radio / Tech Modifications

STANDARD/HEATH Radio Modifications

STANDARD C168A

EXPANDED RF

1. Remove Battery and Antenna.
2. Remove screws and open the case. (Be careful. Do not break flat cables)
3. Locate microprocessor. (see Drawing)
4. Install a DA-113 chip diode in place. (A 1N914 may be used)
5. Reasseble the radio.
6. If required, RESET the microprocessor (see instruction manual)

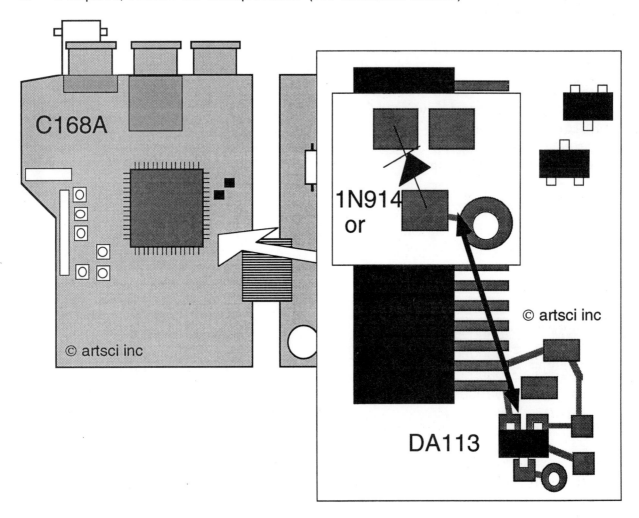

MORE ---

STANDARD C168S

EXPANDED RX / Keyboard

If your radio ends with an "S"

57 - 97 MHz RX AM/FM /
100 - 175 MHz RX AM/FM
213 - 391 MHz RX AM/FM
115 - 174 MHz TX/RX FM

1.   Turn Power on.
2.   Press [ENT]
3.   Press [0], [9].
4.   Press [ENT]
5.   Press and hold [F] then [0].
6.   Press and hold [F] then [ENT].
7.   Press and hold [F] then [0].
8.   Press and hold [F] then [0].
9.   Press and hold [F] then [8].
10. Press [CL]

All Models

DIRECT FREQENCY ENTRY

1.   Press and hold [F] then [0].
2.   Press and hold [F] then [0].
3.   Press [8].

C168 AM / FM mode switch

1.   Press and hold [F] then [0].
2.   Press and hold [F] then [2].

STANDARD C188A

EXPANDED RX

1. Remove Power and Antenna.
2. Remove screws and open case.
3. Locate the microprocessor board
4. Locate QL12 & QL13. (QL13 may already be missing)
5. **Remove QL12 & QL13.** (QL13 may already be missing)
6. Reassemble the radio
7. **Reset Microprocessor** (set mode 8).

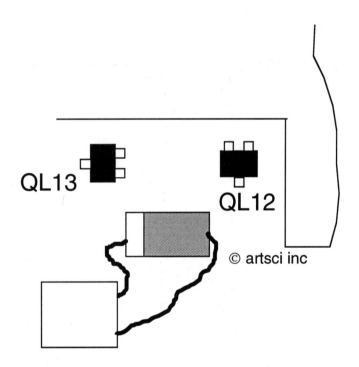

QL13

QL12

© artsci inc

STANDARD C228A

EXPANDED RF

1. Remove Battery and Antenna.
2. Remove two screws from the back case.
3. Remove the four screws from the battery retaining slide.
4. **Insert a 1N914 or DA113 chip diode** in the pictured location.
5. Reassemble the radio.

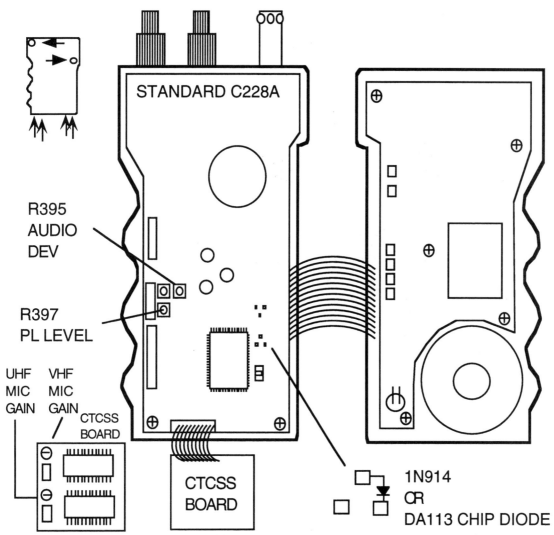

Caution

These modifications have not been tested. The Author, Publisher and all other parties takes NO responsibility or liability for any damage or violation resulting from these modifications. Performing any modification may be a Violation of FCC Rules and will void the warranty of the radio. Use of any modified radio may be a violation of FCC rules. If you have any doubts, DO NOT PERFORM THIS MODIFICATION.

STANDARD C468A

EXPANDED RF

1. Remove Battery and Antenna.
2. Remove screws and open the case. (Be careful. Do not break flat cables)
3. Locate microprocessor. (see Drawing)
4. **Install a DA113 chip diode** in place. (A 1N914 may be used)
5. Reasseble the radio.
6. If required, RESET the microprocessor (see instruction manual)

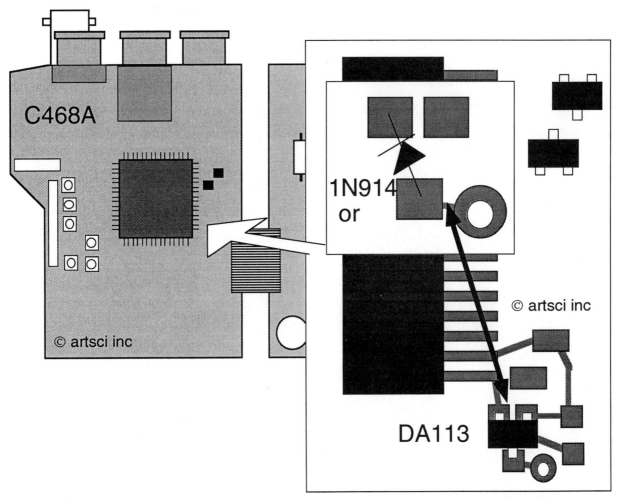

MORE ---

STANDARD C468

EXPANDED RX / Keyboard

If your radio ends with an "S"

340 - 399.995 MHz RX
400 - 474.000 MHZ RX/TX
801 - 980.000 MHz RX

1.   Turn Power on.
2.   Press [ENT]
3.   Press [0], [9].
4.   Press [ENT]
5.   Press and hold [F] then [0].
6.   Press and hold [F] then [ENT].
7.   Press and hold [F] then [0].
8.   Press and hold [F] then [0].
9.   Press and hold [F] then [8].
10.   Press [CL]

All Models

DIRECT FREQENCY ENTRY

1.   Press and hold [F] then [0].
2.   Press and hold [F] then [0].
3.   Press [8].

Caution

STANDARD C488A

EXPANDED RX

1.   Remove Power and Antenna.
2.   Remove screws and open case.
3.   Locate the microprocessor board
4.   Locate QL12 & QL13. (QL13 may already be missing)
5.   **Remove QL12 & QL13.** (QL13 may already be missing)
6.   Reassemble the radio
7.   Reset Microprocessor (set mode 8).

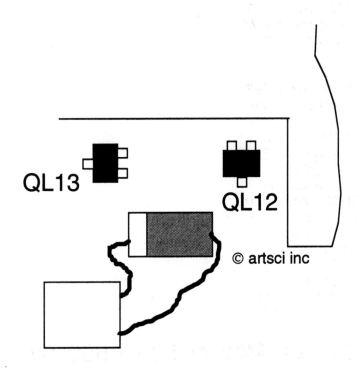

QL13

QL12

© artsci inc

STANDARD C528A

EXPANDED RF / Keyboard / Mars/Cap

1. Turn Power on.
2. Push RESET.
3. Press and hold [FUNCTION] then [0]
4. Press and hold [FUNCTION] then [ENT]
5. Press PTT Briefly.
6. Press [UHF]
7. Press and hold [FUNCTION] then [LAMP]
8. Press and hold [FUNCTION] then [0]
9. Press and hold [FUNCTION] then [CODE]
10. Press and hold [FUNCTION] then [LAMP]
11. Press and hold [FUNCTION] then [3]
12. Press PTT Briefly.
13. Press [VHF]
14. Press and hold [FUNCTION] then [STEP]
15. Select 12.5 KHz. (Use Selector Knob)
16. Press PTT Briefly.
17. Press and hold [FUNCTION] then [8]
18. Press and hold [FUNCTION] then [8]
19. Press and hold [FUNCTION] then [7]
20. Press and hold [FUNCTION] then [7]
21. Press and hold [FUNCTION] then [MS.M]
22. Select 144.9975 MHz (Use Selector Knob)
23. Press and hold [FUNCTION] then [0]
24. Press and hold [FUNCTION] then [ENT]
25. Press PTT Briefly.
26. Press and hold [FUNCTION] then [8]
27. Press and hold [FUNCTION] then [MS.M]

To Receive 300 - 400 Mhz or 800 - 900 MHz

Press [UHF]
Press and hold [FUNCTION] then [SET]
Press and hold [FUNCTION] then [3] to Select Bands

STANDARD C558A

EXPANDED RX

1. Remove Battery and antenna.
2. Locate and remove body screws and open the case.
3. Locate and unsolder the copper plate from the back side of the LCD displat.
4. Locate and remove chip diode D2. (see drawing)
5. **Attach a 1SS301 chip diode** in the vacant D2 position.
6. Reassebmle the radio.
7. Reset the microprocessor, if required.

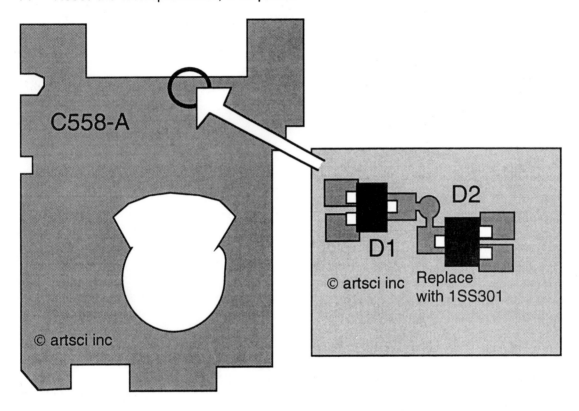

Performance Report

Radio _____ Date _____

Owner : Name _____
            Address _____
            City _____ St. ____ Zip _____
            Phone (____) _____ - _____

Description	Before	After
Power out (Low)	_____ Watts	_____ Watts
Power out (High)	_____ Watts	_____ Watts
Frequency Error (Simplex)	_____ Hz	_____ Hz
Frequency Error (Offset)	_____ Hz	_____ Hz
Receive Sensitivity (Mid-band)	_____ uv	_____ uv
Receive Sensitivity (____MHz)	_____ uv	_____ uv
Receive Sensitivity (____MHz)	_____ uv	_____ uv
PL Deviation	_____ Hz	_____ Hz
DTMF Deviation	_____ KHz	_____ KHz
Audio Deviation	_____ KHz	_____ KHz
Lowest usable Freq @ .5 Pwr	_____ MHz	_____ MHz
Highest usable Freq @ .5 Pwr	_____ MHz	_____ MHz

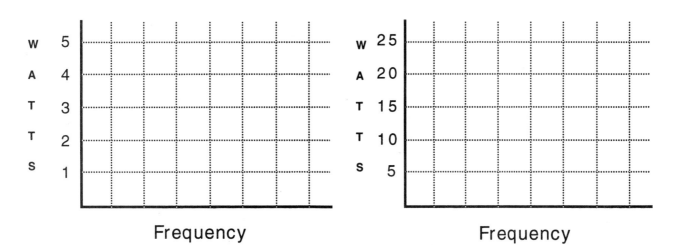

W A T T S 5 4 3 2 1   Frequency   W A T T S 25 20 15 10 5   Frequency

STANDARD C5608DA

EXPANDED RF

1. Remove power and antenna.
2. **Remove 0 ohm resistors** near the microprocessor.
   Specific data:

   RL169 "H" symbol   400-469.996 MHz TX
                                250-499.995 MHz RX

   RL70 "D" symbol   130-173.995 MHz TX
                                100-199.995 MHz RX

3. Reassemble the radio.
4. Reset the microprocessor (if required)

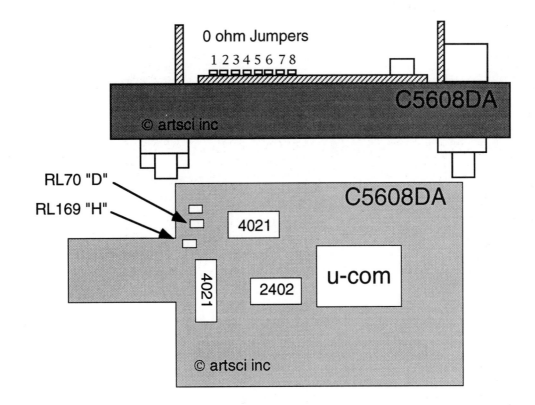

STANDARD C5608DA

800 MHz Modification

1. Remove power and antenna.
2. Remove covers
3. Remove black tape patch under the VHF antenna connector.
4. Remove cover from transmitter (5 screws)
5. Remove screws securing the red and black power wires.
6. **Solder attach the new antenna coax** as shown.
7. Secure the coax using wire ties or other method.
8. Replace the power cable screws.
9. Replace the covers.

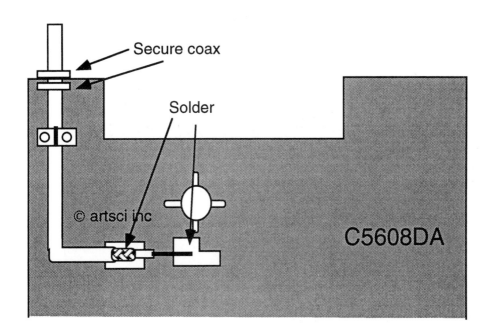

800 MHz activation:

Select 440 as the main band.
Press [UP] button while pressing the rotary switch
Press [UP] button while pressing the [FUNCTION] button.
To Return to 440 - Press [DOWN] while pressing [FUNCTION] button.

STANDARD HT TNC Hookup

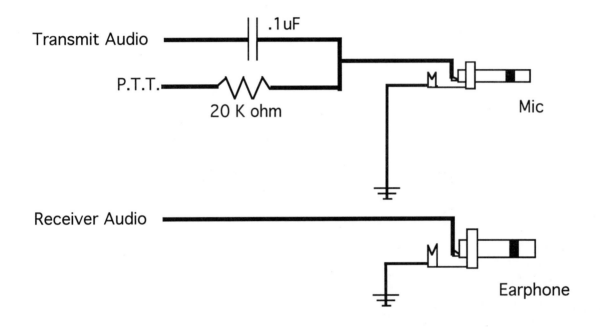

Transmit Audio

.1uF

P.T.T.

20 K ohm

M

Mic

Receiver Audio

M

Earphone

HEATH H-2 Mini HT

EXPANDED RF
130 - 169.995 MHz

1. Remove battery and Antenna.
2. Remove 2 lower screws from the battery plate.
3. Remove 2 screws securing thr front & back cases.
4. Locate Q12 Position. (find point A and B)
5. **Solder a diode (1N914 or eq.) from point A to point B**
   Cathode to point A, Anode to Point B.
6. Reassemble the radio.
7. Reset the microprocessor.

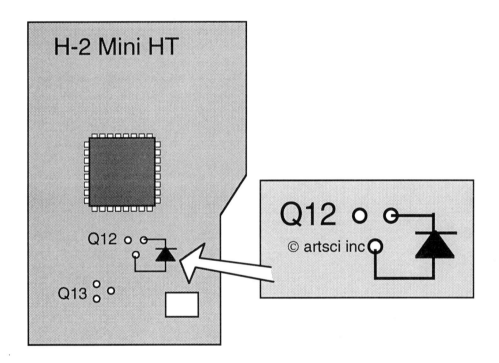

Caution

These modifications have not been tested. The Author, Publisher and all other parties takes NO responsibility or liability for any damage or violation resulting from these modifications. Performing any modification may be a Violation of FCC Rules and will void the warranty of the radio. Use of any modified radio may be a violation of FCC rules. If you have any doubts, DO NOT PERFORM THIS MODIFICATION.
PHOTOCOPIES OF THESE PAGES ARE A VIOLATION OF COPYRIGHT LAW.
© artsci inc. all rights reserved. (818) 843-4080 Fax: (818) 846-2298

Performance Report

Radio _____ Date _____

Owner : Name _____
         Address _____
         City _____ St. _____ Zip _____
         Phone (_____) _____ - _____

Description	Before	After
Power out (Low)	_____ Watts	_____ Watts
Power out (High)	_____ Watts	_____ Watts
Frequency Error (Simplex)	_____ Hz	_____ Hz
Frequency Error (Offset)	_____ Hz	_____ Hz
Receive Sensitivity (Mid-band)	_____ uv	_____ uv
Receive Sensitivity (____MHz)	_____ uv	_____ uv
Receive Sensitivity (____MHz)	_____ uv	_____ uv
PL Deviation	_____ Hz	_____ Hz
DTMF Deviation	_____ KHz	_____ KHz
Audio Deviation	_____ KHz	_____ KHz
Lowest usable Freq @ .5 Pwr	_____ MHz	_____ MHz
Highest usable Freq @ .5 Pwr	_____ MHz	_____ MHz

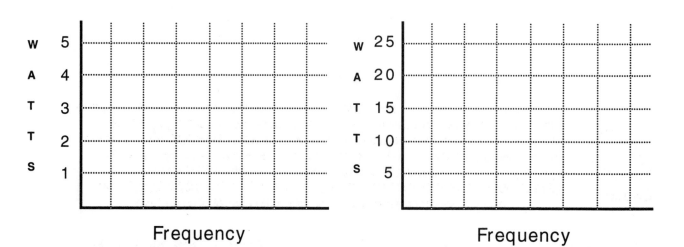

Frequency   Frequency

HEATH H4-HT Twin Band

EXPANDED RF

1. Remove battery and Antenna.
2. Remove 2 lower screws from the battery plate.
3. Remove 2 screws securing thr front & back cases.
4. Locate Q106 Position. (find point A and B)
5. Solder a diode (1N914 or eq.) from point A to point B
    Cathode to point A, Anode to Point B.
6. Reassemble the radio.

HEATH HW-24

EXPANDED RF

1. Remove power and Antenna.
2. Remove the wire mounting stand.
3. Remove the five screws that hold the bottom cover.
4. Remove the bottom plate being careful to unplug the speaker as you remove it.
5. Locate and cut the lead of QD22 (2 meter RX Mod)
6. Locate and cut the lead of QD24 (440 RX Mod)
7. Locate Chip Diode QD23 on front panel board.
8. **Cut leads to both bottom leads of QD23**. (note it may be required to remove the front panel from the body of the radio.)
9. Reassemble the radio (see next step)
10. **Reset the Radio.** (short the Reset pins with a wire or screw driver.)

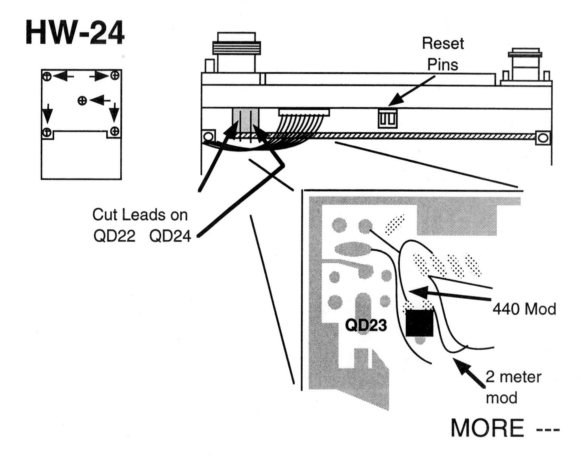

HW-24

Reset Pins

Cut Leads on
QD22 QD24

QD23

440 Mod

2 meter mod

MORE ---

HEATH HW-24

EXPANDED RF / Keyboard / Mars/Cap

1.   Turn Power on.
2.   Push RESET.
3.   Press and hold [FUNCTION] then [0]
4.   Press and hold [FUNCTION] then [ENT]
5.   Press PTT Briefly.
6.   Press [UHF]
7.   Press and hold [FUNCTION] then [LAMP]
8.   Press and hold [FUNCTION] then [0]
9.   Press and hold [FUNCTION] then [CODE]
10.   Press and hold [FUNCTION] then [LAMP]
11.   Press and hold [FUNCTION] then [3]
12.   Press PTT Briefly.
13.   Press [VHF]
14.   Press and hold [FUNCTION] then [STEP]
15.   Select 12.5 KHz. (Use Selectror Knob)
16.   Press PTT Briefly.
17.   Press and hold [FUNCTION] then [8]
18.   Press and hold [FUNCTION] then [8]
19.   Press and hold [FUNCTION] then [7]
20.   Press and hold [FUNCTION] then [7]
21.   Press and hold [FUNCTION] then [MS.M]
22.   Select 144.9975 MHz (Use Selector Knob)
23.   Press and hold [FUNCTION] then [0]
24.   Press and hold [FUNCTION] then [ENT]
25.   Press PTT Briefly.
26.   Press and hold [FUNCTION] then [8]
27.   Press and hold [FUNCTION] then [MS.M]

To Receive 300 - 400 Mhz or 800 - 900 MHz

Press [UHF]
Press and hold [FUNCTION] then [SET]
Press and hold [FUNCTION] then [3] to Select Bands

HEATH SB-1400

EXPANDED RF

1. Remove power and Antenna.
2. Remoce screws and ope the case.
3. Locate the BROWN jumper wire on the display unit.
4. **Cut the BROWN jumper wire.**
5. Reassemble the radio.
6. Reset the microprocessor.
   (Set VFO at 12.3456 MHz, Turn power of and on again)

Radio / Tech Modifications
YAESU Radio Modifications

Radio / Tech Modifications
YAESU Radio Modifications

YAESU FL-7000

EXPANDED RF 24.5 MHz & 28.0 MHz Band

1. Remove Power cable and all other cables.
2. Remove 4 screws from the top cover.
3. Remove the top cover and the right and left panels.
4. Remove 4 screws from the power combiner unit and remove screen plate.
5. **Locate Switch SO1 on the CPU unit and set it to the off position.** (A small screwdriver can be used to reach the switch.)
6. Reassemble the unit.

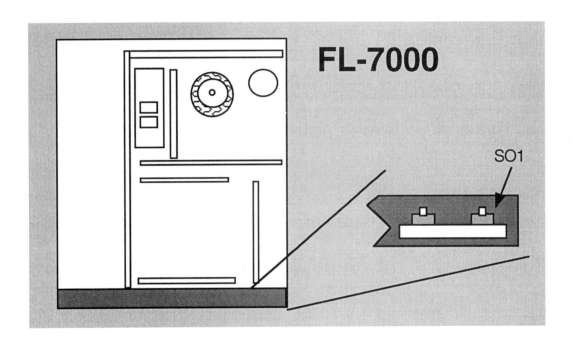

Performance Report

Radio _____ Date _____

Owner : Name _____
          Address _____
          City _____ St. ____ Zip _____
          Phone (_____) ____-_____

Description	Before	After
Power out (Low)	_____ Watts	_____ Watts
Power out (High)	_____ Watts	_____ Watts
Frequency Error (Simplex)	_____ Hz	_____ Hz
Frequency Error (Offset)	_____ Hz	_____ Hz
Receive Sensitivity (Mid-band)	_____ uv	_____ uv
Receive Sensitivity (____MHz)	_____ uv	_____ uv
Receive Sensitivity (____MHz)	_____ uv	_____ uv
PL Deviation	_____ Hz	_____ Hz
DTMF Deviation	_____ KHz	_____ KHz
Audio Deviation	_____ KHz	_____ KHz
Lowest usable Freq @ .5 Pwr	_____ MHz	_____ MHz
Highest usable Freq @ .5 Pwr	_____ MHz	_____ MHz

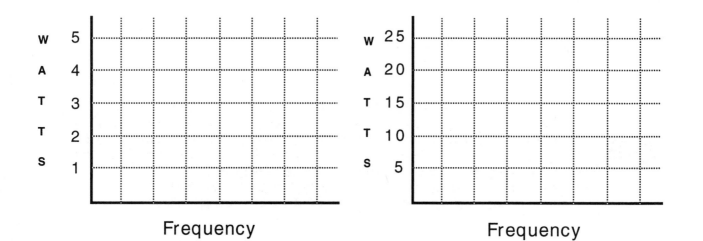

W A T T S   5 4 3 2 1   Frequency

W A T T S   25 20 15 10 5   Frequency

YAESU FT-23R

EXPANDED RF & ALIGNMENT CONTROLS

1. Remove Battery and Antenna.
2. Remove control knobs, screws, top panel, battery mounting track & body screws and open Radio
3. **Remove solder bridge from Pad # 7**
4. Reassemble radio.

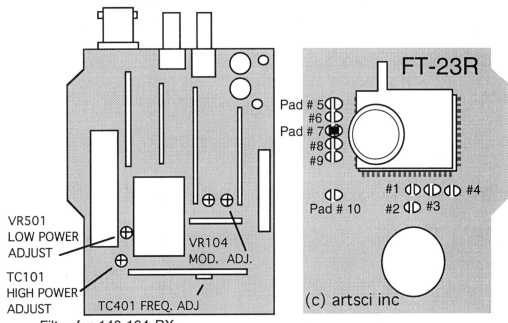

Pad #1   Filter for 140-164 RX
Pad #2   Filter for 164-?? RX
Pad #3&4   Step selection 20 or 25 KHz   3&4 unsoldered = 10 KHz step
Pad #5   5 MHz offset
Pad #6   1.6 MHz offset   5&6 unsoldered = 600 kHz offset
Pad #7,8&9   Band selections
Pad #10   Unknown

Range :   RX 140 MHz - 163.995 MHz
        TX 140 MHz - 163.995 MHz

YAESU FT-26

EXPANDED RF
New Range: 135 - 174 MHz

1. Remove Battery and Antenna.
2. Remove the 4 screws holding the battery track.
3. Remove the 2 screws in the back case.
4. Carefully separate the front cover.
5. **Locate and remove solder on Jumper pad 10.** (on control board)

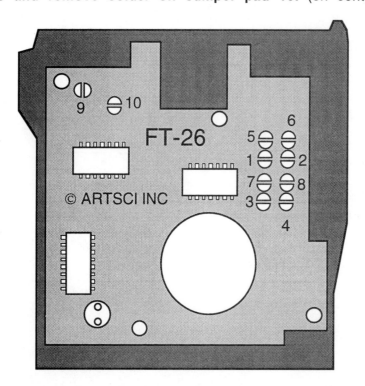

6. **Solder jump pads 1, 3, 7 and 8.**
7. Reassemble the radio.
8. Turn radio on and each channel indicator will blink.
9. Enter the following frequencies. (use the [F] & up arrow keys)

CH. 1	135.000	Press [D/MR] Lower Rx limit
CH. 2	174.000	Press [D/MR] Upper Rx limit
CH. 3	135.000	Press [D/MR] Lower Tx limit
CH. 4	174.000	Press [D/MR] Upper Tx limit

MORE ---

YAESU FT-26
ALIGNMENT POINTS

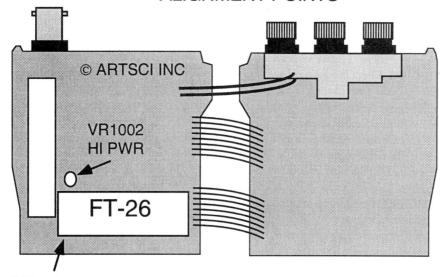

© ARTSCI INC

VR1002
HI PWR

FT-26

VR1001
DEV. ADJUSTMENT

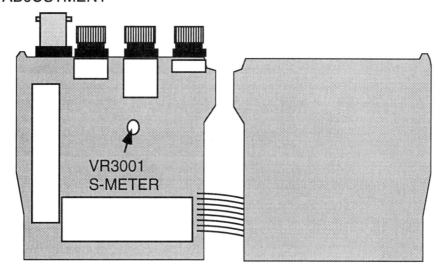

VR3001
S-METER

RESET COMMANDS:

Soft RESET   Press and hold [T] & [REV] and turn power on.

Master RESET   Press and hold [D/MR] & [T] & [REV] and turn radio on.
Then enter band Limits above

YAESU FT-33R

EXPANDED RF & ALIGNMENT CONTROLS

1. Remove Battery and Antenna.
2. Remove control knobs, screws,top panel, battery mounting track & body screws and open Radio
3. For display 220-550 MHz **Pads 7,8 and 9 are open**
   For display 50-300 MHz Pads 8 and 9 are open and 7 is bridged
4. Reassemble radio.

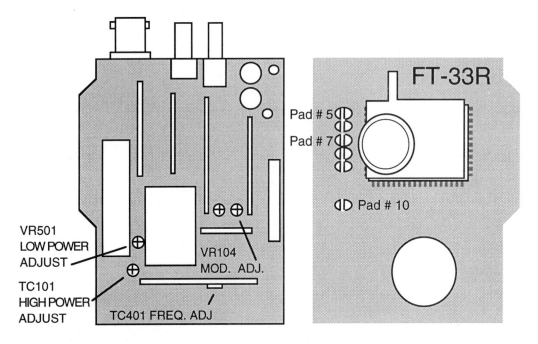

Note: The exact TX and RX range is determined by the coils and other circuitry in the radio.

YAESU FT-73R

ALIGNMENT CONTROLS

1. Remove Battery and Antenna.
2. Remove control knobs, screws,top panel, battery mounting track & body screws and open Radio
3. Make adjustments.
4. Reassemble the radio.

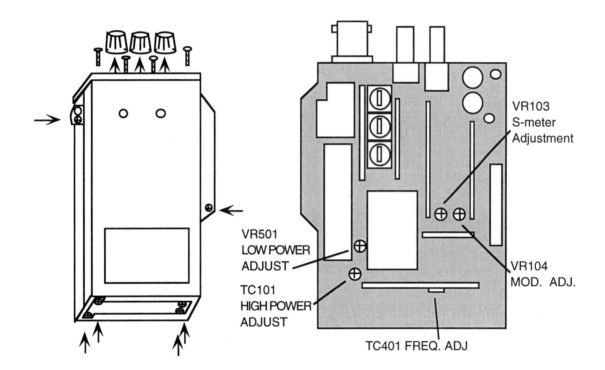

VR103
S-meter
Adjustment

VR501
LOW POWER
ADJUST

VR104
MOD. ADJ.

TC101
HIGH POWER
ADJUST

TC401 FREQ. ADJ

YAESU FT-76
EXPANDED RF

1. Remove Battery and Antenna.
2. Remove the 4 screws holding the battery track.
3. Remove the 2 screws in the back case.
4. Carefully separate the front cover.
5. **Locate and remove solder on Jumper pads 4 and 7.** (on control board)
6. **Solder jump pads 1, 3, 5, 8, 9 and 10** (old mod had pad 4 in place of 5)

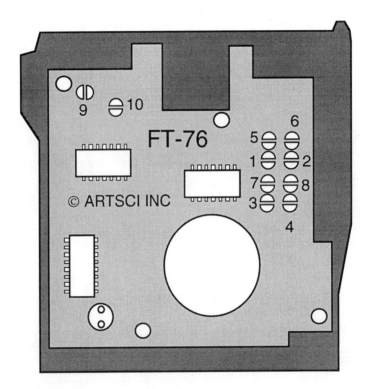

7. Reassemble the radio.
8. Turn radio on and each channel indicator will blink.
9. **Enter the following frequencies.** (use the [F] & up arrow keys)

CH. 1	400.000	Press [D/MR] Lower Rx limit
CH. 2	485.000	Press [D/MR] Upper Rx limit
CH. 3	415.000	Press [D/MR] Lower Tx limit
CH. 4	470.000	Press [D/MR] Upper Tx limit

New Range: 400 - 485 MHz RX, 415 - 470 MHz TX

MORE ---

YAESU FT-76
ALIGNMENT POINTS

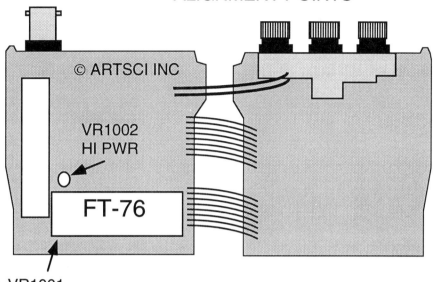

VR1002
HI PWR

FT-76

VR1001
DEV. ADJUSTMENT

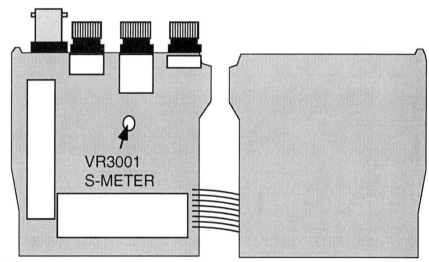

VR3001
S-METER

RESET COMMANDS:

Soft RESET   Press and hold [T] & [REV] and turn power on.

Master RESET   Press and hold [D/MR] & [T] & [REV] and turn radio on.
Then enter band Limits above

YAESU FT-209
ALIGNMENT POINTS / Untested Mods

1. Remove battery and antenna.
2. Remove battery screws, belt clip screws and side strap screws.
3. Remove black trim on sides of the radio.
4. Remove the two side screws and slide the u-shaped back cover off.
5. Remove the four tiny phillips screws holding the front panel on.
6. Fold panel to the right to open the radio.

Untested out of band mod #1: **Jumper pads 1,7,9,10 & 13.**
Untested out of band mod #2: **Jumper pads 7,9,10,11& 13.**
     Factory default is pads 1,9 & 13.

7. Locate alignment pots. Make adjustments
8. Reassemble the radio.
9. Reset the microprocessor (If desired)
10. Enter 1440 [D], 1480 [D], 1440 [D], 1480 [D], 0600 [SHIFT]
    Note: RX range of 144.0 - 148.0 MHz and TX range of 144.0 - 148.0 MHz

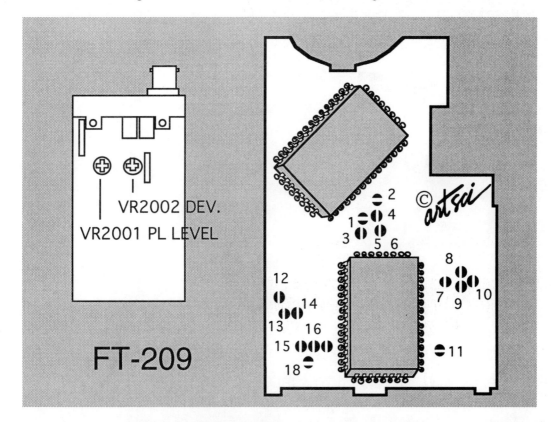

YAESU FT-211

EXPANDED RF & ALIGNMENT CONTROLS

1. Remove five screws from the top cover and remove the cover.
2. Remove five screws from the bottom cover and remove the cover.
3. Unplug the speaker.
4. Remove the four screws holding the front panel.
5. Locate jumper pad number 7.
6. **Solder bridge pad number 7.**
7. Locate the reset pins (Located on the front panel and clearly marked).
8. **Short the reset pins together for one second.**
9. Reassemble the radio.

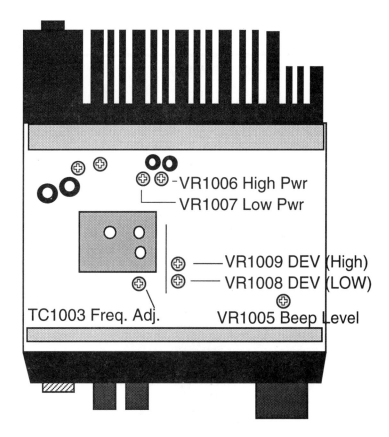

YAESU FT-212
EXPANDED RF
Auto Repeater offset is lost

1. Unplug the DC power cable from the radio.
2. Remove the top and bottom covers.
3. Remove the speaker.
4. Remove the knobs and nuts from the front panel.
5. Remove the three screws from the control unit.
6. Remove the Control unit from the front panel.
7. **Locate & remove solder from pad #1 on control unit.**
8. **Locate & solder jumper Pads 3,4,11 and 14.**
9. Replace the control unit on the front panel.
10. **Reset the microprocessor.** (using a jumper short D09 on the control unit to ground on the radio. Do not apply power).
11. Reassemble the radio. Replace knobs, screws etc.
12. Apply DC power and turn radio on.
13. Press [MHz] & use the control knob to enter 140 and press [D/MR]. (lower limit)
14. Press [MHz] and use knob to enter 174 and press [D/MR]. (upper limit)
15. Press [F] and then [RPT] button. use the control knob to enter 0.600. Press the [RPT] button.
    Note: New range 140 - 164 MHz

FT-212

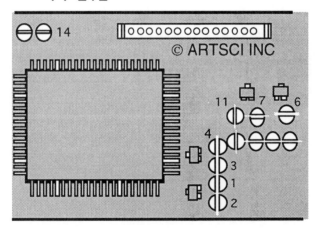

MORE ---

YAESU FT-212

ALIGNMENT CONTROLS

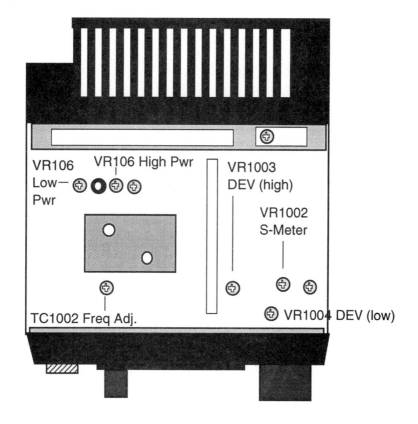

VR106
Low—
Pwr

VR106 High Pwr

VR1003
DEV (high)

VR1002
S-Meter

TC1002 Freq Adj.

VR1004 DEV (low)

YAESU FT-227R

EXPANDED RF & ALIGNMENT CONTROLS

1.   Unplug the power from the radio.
2.   Open radio and locate the PLL CONT. UNIT.
3.   **Remove D701 and D702.** Do not place in a jumper.
4.   **Locate Q712 (MC14028B), and break the connection to Pin 6. (Blue wire)**
5.   **Connect pin 1 of Q711 (red wire) to ground.**
6.   Reassemble radio

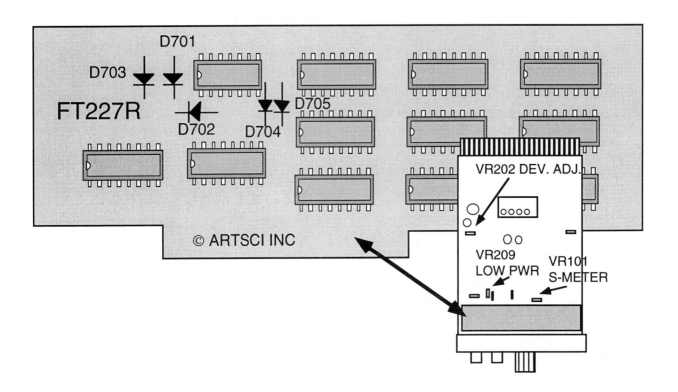

Note: Automatic repeater offset is lost.

TX Range   143.990 MHz - 149.000 MHz

YAESU FT-290 MKII

EXPANDED RF & ALIGNMENT CONTROL

1. Unplug the power from the radio.
2. Open radio and located SW Unit. The SW unit is located on the front panel, behind the display.
3. Locate components D01, D03, R02 & R03 See drawing.
4. **Remove or Install the components per table 1.**
5. Reassemble the radio.

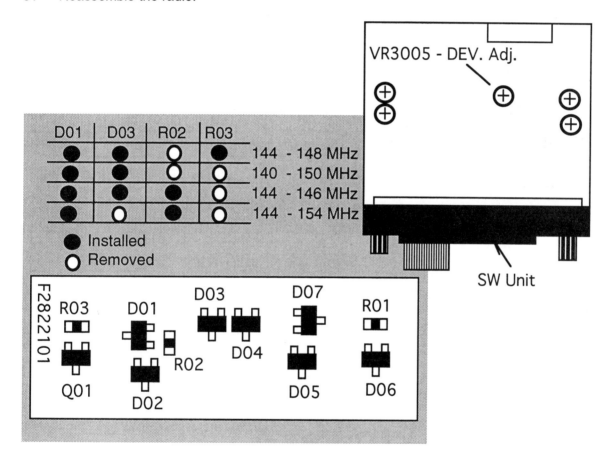

D01	D03	R02	R03	
●	●	○	●	144 - 148 MHz
●	●	○	○	140 - 150 MHz
●	●	●	○	144 - 146 MHz
●	○	●	○	144 - 154 MHz

● Installed
○ Removed

VR3005 - DEV. Adj.

SW Unit

YAESU FT-311

EXPANDED RF & ALIGNMENT CONTROLS

1. Remove five screws from the top cover and remove the cover.
2. Remove five screws from the bottom cover and remove the cover.
3. Unplug the speaker.
4. Remove the four screws holding the front panel.
5. Locate jumper pad number 7.
6. **Solder bridge pad number 7.**
7. Locate the reset pins (Located on the front panel and clearly marked).
8. **Short the reset pins together for one second.**
9. Reassemble the radio.

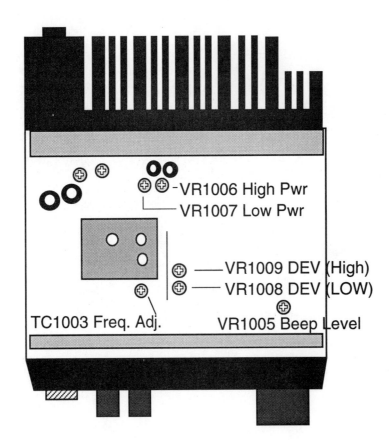

YAESU FT-411 E

EXPANDED RF & ALIGNMENT CONTROLS
(disables automatic repeater shift)

1. Remove Battery and Antenna.
2. Remove control knobs, screws,top panel & body screws and open Radio
3. **Remove solder bridge from Pad # 2**
4. **Place solder Bridge on Pad # 3**
5. Reassemble Radio
6. **Reset Microprocessor.**
   (Press and hold [MR], [2] & [VFO] and turn radio on then off)
   (Press and hold both up and down keys and turn power on)
7. Enter the following: 1200 [VFO]   1740 [VFO]   1400 [VFO]   1740 [VFO]
8. Press [Function] & [7] to change channel step.

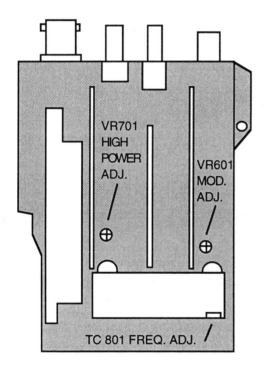

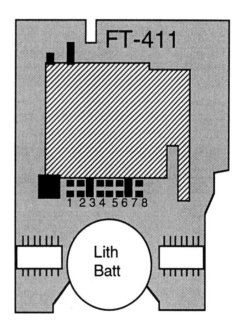

RANGE :   RX 120 MHz - 174 MHz
            TX 140 Mhz - 174 MHz

YAESU FT-415
EXPANDED RF
Range : 120 - 174 MHz RX,   135 - 174 MHz TX

1.   Remove Battery and Antenna.
2.   Remove the four screws holding the battery track in place.
3.   Remove the two black screws holding the rear case in place.
4.   Carefully open the front cover from the radio.
5.   **Locate and solder jumper pads 5 & 7.**   Pads 3 and 9 are already jumpered.
     (Jumper pads 1 & 10 for 1750 Hz Tone Burst operation)

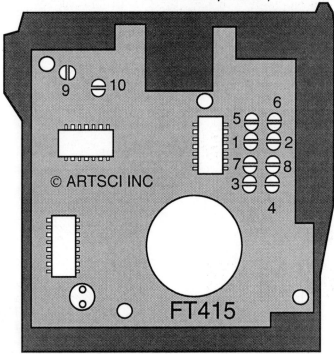

6.   Carefully replace the front cover and replace the two black screws.
7.   Replace the battery track and the four screws.
8.   **Reset the microprocessor.**
     Press and hold [MR], [2] and [VFO] and turn the radio on.
9.   The radio display will cycle orderly thru the memory channels.
     Enter the following band limits:

        Ch. 1   Enter 120.00 and then press [VFO]   (Rx low limit)
        Ch. 2   Enter 174.00 and then press [VFO]   (Rx high limit)
        Ch. 1   Enter 135.00 and then press [VFO]   (Tx low limit)
        Ch. 1   Enter 174.00 and then press [VFO]   (Tx high limit)

MORE---

YAESU FT-415
EXPANDED RF & ALIGNMENT POINTS

10. Press [F] [7] and select 5 kHz channel spacing in each VFO.

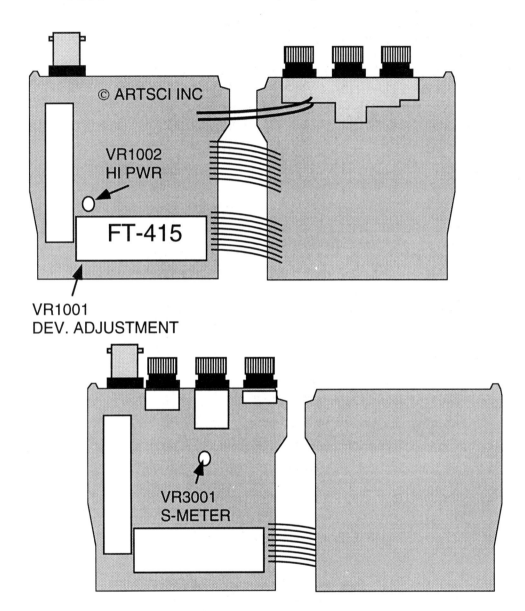

Master Reset Command: Press and hold [MR] & [2] & [VFO] and turn power on, then enter new limits

YAESU FT-416
EXPANDED RF

Range : 123 - 174 MHz RX,   135 - 174 MHz TX

1. Remove Battery and Antenna.
2. Remove the four screws holding the battery track in place.
3. Remove the two black screws holding the rear case in place.
4. Carefully open the front cover from the radio.
5. **Locate and solder jumper pads 5 & 7.**   Pads 3 and 9 are already jumpered.
   (Jumper pads 1 & 10 for 1750 Hz Tone Burst operation)

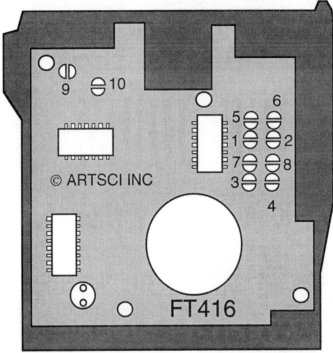

6. Carefully replace the front cover and replace the two black screws.
7. Replace the battery track and the four screws.
8. **Reset the microprocessor.**
   Press and hold [MR], [2] and [VFO] and turn the radio on.
9. The radio display will cycle orderly thru the memory channels.
   Enter the following band limits:

   Ch. 1   Enter 120.00 and then press [VFO]   (Rx low limit)
   Ch. 2   Enter 174.00 and then press [VFO]   (Rx high limit)
   Ch. 1   Enter 135.00 and then press [VFO]   (Tx low limit)
   Ch. 1   Enter 174.00 and then press [VFO]   (Tx high limit)

MORE---

YAESU FT-416
EXPANDED RF & ALIGNMENT POINTS

10. Press [F] [7] and select 5 kHz channel spacing in each VFO.

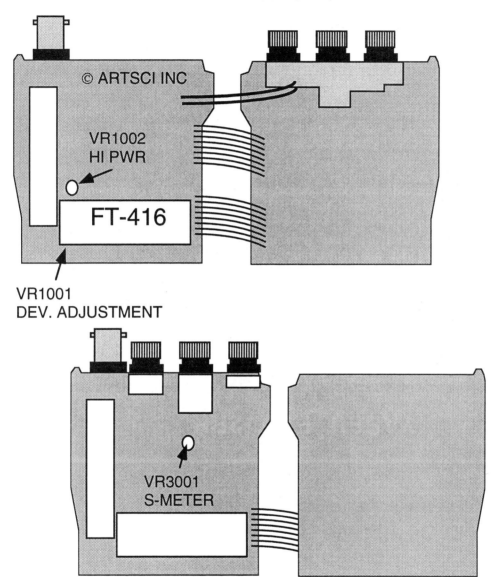

Master Reset Command: Press and hold [MR] & [2] & [VFO] and turn power on, then enter new limits

YAESU FT-470
EXPANDED RF & ALIGNMENT POINTS

1. Remove Battery and Antenna.
2. Remove control knobs, screws, top panel & body screws and open Radio
3. Locate the lithium battery.
4. Carefully unsolder the lithium battery and lift it to expose resistor position .
5. **Solder a Jumper or 0 ohm resistor(or jumper) in the empty R69 position.**
6. OPTIONAL- Crossband Half Duplex mod. Place a jumper wire from pin 4 & 14 of the flat cable wire connecting the front and back panels. This will use the ON AIR signal to mute the AUDIO CNTL line, muting the other band while transmitting.
7. Solder the lithium battery back in place.
8. Reassemble the radio.

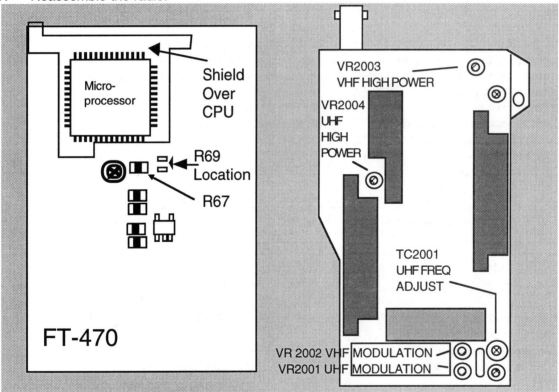

Range   140 MHz - 174 MHz
Note: Freq. expansion is possible using the keyboard only:
    1. Press and hold [MR] and [VFO] Buttons and turn radio on
    2. Release buttons and turn radio off. (Stop here for normal operation)
    3. Press and hold [up] and [down] buttons and turn radio on.
       Range: 140-150 MHz TX/RX and 430-450 MHz TX/RX

MORE ---

YAESU FT-470
EXPANDED 430-500 MHz RX/ Hyperscan Mod

The following procedure utilizes the "U" memory location to store the upper limit for the UHF reception. A high UHF frequency (ie 470 MHz) must always be stored in the "U" memory for the expanded UHF reception to work.

1. Program 450.00 MHz simplex.
2. Press [F/M] and then [RPT].
3. Enter 0000 into the keypad.
4. Turn the radio off and turn back on.
5. Press [RPT] twice for a + (plus) offset.
6. Press the [REV] button. (The display should now be 1450 MHz)
7. Press [Function] and then [Down Arrow] to drop the frequency down 1 MHz at a time until the display reads 500 MHz.
8. Press and hold the [F/M] key until your hear two beeps.
9. Rotate the dial knob until the "U" memory channel is displayed.
10. Press the [Function] key to store the frequency in memory.
11. Press [Function] and then [Down Arrow] to drop the frequency down 1 MHz at a time until the display reads 450 MHz.
12. Press and hold the [F/M] key until your hear two beeps.
13. Rotate the dial knob until the "L" memory channel is displayed.
14. Press the [Function] key to store the frequency in memory.
*** **Stop here for 440 - 470 Coverage.**
15. Turn radio off and on and select the "U" memory channel.
16. Press [MR] and then [RPT]
17. Press the PTT button 3 times. The display should read 070.00 MHz
18. Press [Function] and then [Up Arrow] to increase the frequency up1 MHz at a time until the display reads 400 MHz.
19. Press and hold the [Function] key until your hear two beeps.
20. Rotate the dial knob until the "L" memory channel is displayed.
21. Press the [Function] key to store the frequency in memory.

To receive a desired UHF frequency, you must use the following steps:
1. Select the "U" memory channel.
2. Press the [MR] key to enter the "MEMORY TUNE" mode.
3. Use the [arrow] keys or Dial Knob to select the desired frequency.
4. Store the selected in any memory channel, except memory channel "U" & L

Hyperscan Modification:
1. Select the "ALT mode by pressing [F] and [ALT]
2. Press the [UP] or [DOWN] arrow.
3. When the scan stops, Press [F] and then [VFO].
4. Press the [UP] or [DOWN] arrow. (HYPERSCAN MODE)
5. Press [F] and [ALT] to stop scan mode.

YAESU FT-530

EXPANDED RF

1. Remove battery and antenna.
2. Locate and remove the 4 screws on the bottom battery track.
3. Locate and remove the 4 black screws on the rear case.
4. Carefully open the front cover and open the radio.
5. Note location of white paper insulator and remove it. (Dont throw away)
6. **Locate jumpers location J13 and remove solder jumper.
   DO NOT DO BOTH JUMPER 13 & 15.**
7. Replace the paper insulator making sure the ground tabs slide through insulator
8. Close radio being careful not to pinch any wires.
9. Replace all screws.
10. Replace battery and antenna.
11. **Press and hold both [UP] & [DOWN] arrow buttons and turn power on..**

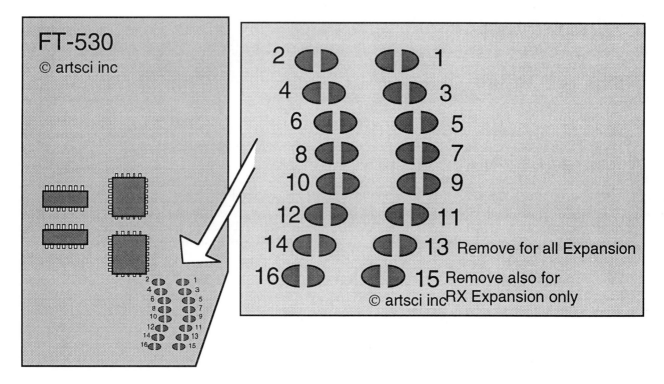

TONE BURST - Jumper Pad # 12.

Stock Pads Soldered: 1, 3, 6, 8, 11, 13, 15

YAESU FT-650

EXPANDED RF

1. Turn the radio off.
2. Press and hold {VFO] & [MR] and turn on the radio.

New Range: 24- 56 MHz

Repeat the step above to return to Normal settings

YAESU FT-709
ALIGNMENT POINTS / Untested Mods

1. Remove battery and antenna.
2. Remove battery screws, belt clip screws and side strap screws.
3. Remove black trim on sides of the radio.
4. Remove the two side screws and slide the u-shaped back cover off.
5. Remove the four tiny phillips screws holding the front panel on.
6. The ground jumper on the left side needs to be unsoldered.
7. Fold panel to the right to open the radio

Untested out of band mod #1: **Jumper pads 1,7,9,10, 13 & 16.**
Untested out of band mod #2: **Jumper pads 7,9,10,1, 13 & 16.**

8. Locate alignment pots. Make adjustments.
9. Reassemble the radio.
10. Reset the microprocessor. (If desired)
11. On FT-709 enter 4400 [D], 4490 [D], 4400 [D], 4490 [D]. 5000 [SHIFT]
    Note: RX range of 440.0 - 449.0 MHz and TX range of 440.0 - 449.0 MHz

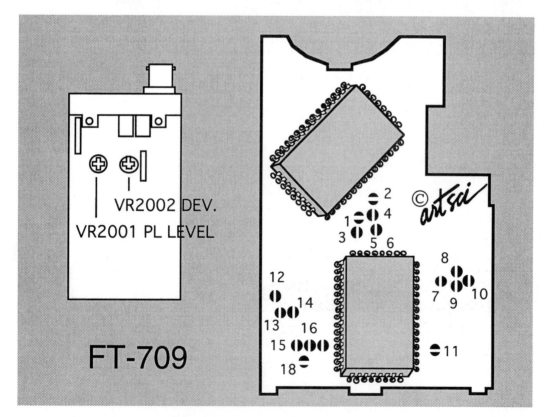

YAESU FT-711

EXPANDED RF & ALIGNMENT CONTROLS

1. Remove five screws from the top cover and remove the cover.
2. Remove five screws from the bottom cover and remove the cover.
3. Unplug the speaker.
4. Remove the four screws holding the front panel.
5. Locate jumper pad number 7.
6. **Solder bridge pad number 7.**
7. Locate the reset pins (Located on the front panel and clearly marked).
8. Short the reset pins together for one second.
9. Reassemble the radio.

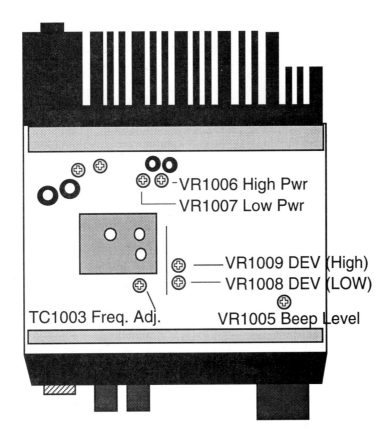

YAESU FT-712RH

EXPANDED RF

1. Unplug the DC power cable from the radio.
2. Remove the top and bottom covers.
3. Remove the speaker.
4. Remove the knobs and nuts from the front panel.
5. Remove the three screws from the control unit.
6. Remove the Control unit from the front panel.
7. **Remove solder from pad #1 and Pad #2 on control unit.**
8. **Solder jumper Pads 4 and 14.** Pads 3,4,5,7,11 and 14 will be bridged
9. Replace the control unit on the front panel.
10. **Reset the microprocessor.** (using a jumper short D09 on the control unit to ground on the radio. Do not apply power).
11. Apply DC power and turn radio on.
12. Press [MR] & use the control knob to enter 430 and press [D/MR]. (lower limit)
13. Press [MR] and use knob to enter 501 and press [D/MR]. (upper limit)
14. Press [F] and then [RPT] button. use the control knob to enter 5.000. Press the [RPT] button.

FT-712

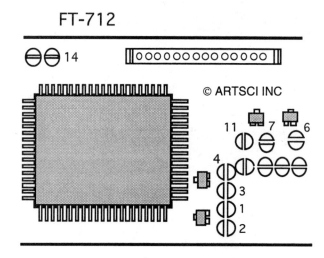

© ARTSCI INC

RANGE: 430 MHz - 465 MHz

MORE ---

YAESU FT-712RH

ALIGNMENT CONTROLS

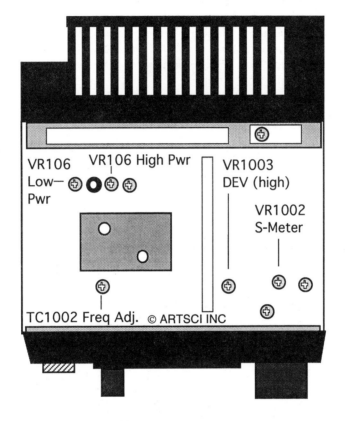

VR106
Low—
Pwr

VR106 High Pwr

VR1003
DEV (high)

VR1002
S-Meter

TC1002 Freq Adj.   © ARTSCI INC

YAESU FT-727
(no 12.5 KHz steps in 440 band)

EXPANDED RF & ALIGNMENT CONTROLS

1. Remove Battery
2. Turn off the Battery backup switch. (located on the bottom of the radio)
3. Wait 10 Seconds and Turn the switch back on
4. Replace battery
5. Turn Radio ON. (Display should go blank, if not redo steps 1-4)
6. Enter the following: 001111 (note: factory setting is 443300)
7. Reset the VHF & UHF offsets.
   Select VHF then Press [F] then the [Shift] button.
   Enter 0600 then [D]
   Select UHF then Press [F] then the [Shift] button.
   Enter 5000 then [D]

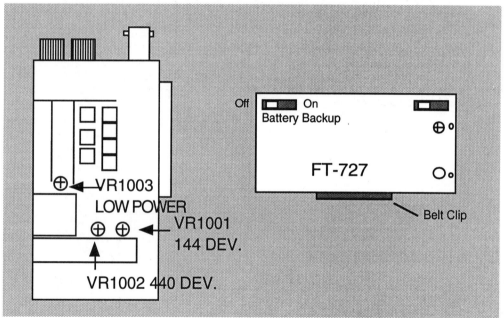

PLL alignment for out of band

1. Remove battery, and belt clip
2. Remove battery track screws
3. Remove rear cover
4. Install the battery track.
5. Turn radio on & enter desired frequency
6. Adjust L01 (black slug) in VCO unit until the on air lamp is lit (red light)
   (L01 core, turn counter-clock wise)
7. Reassemble the radio.

YAESU FT-736R

EXPANDED RF

1. Unplug the power from the radio.
2. Open the radio and locate the 144 MHz main unit.
3. Locate diodes D24, D25, D26 and D27 See drawing.
4. **Remove or Install the diodes per table 1.**
5. Reassemble radio.

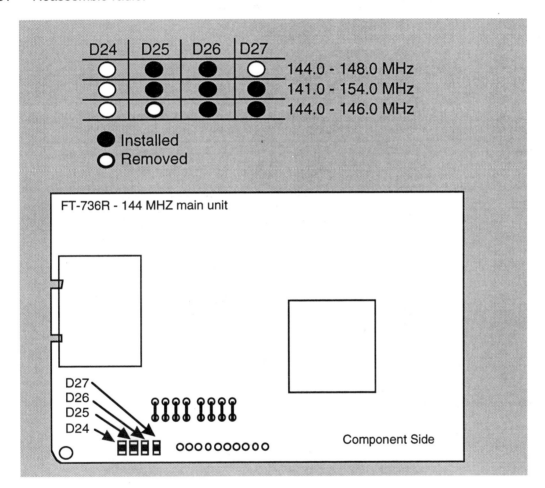

D24	D25	D26	D27	
○	●	●	○	144.0 - 148.0 MHz
○	●	●	●	141.0 - 154.0 MHz
○	○	●	●	144.0 - 146.0 MHz

● Installed
○ Removed

FT-736R - 144 MHZ main unit

D27
D26
D25
D24

Component Side

YAESU FT-747

EXPANDED RF

1. Unplug the DC power cable from the radio
2. Remove the top cover (see instruction manual page 23)
3. **Remove or cut the BROWN jumper wire** on the display unit. See Drawing
4. Reconnect the power cable and turn the radio on
5. Set the VFO dial to 12.3456 MHz
6. Turn power off and then back on again.
7. Turn power off and reassemble radio. (don't pinch any wires)

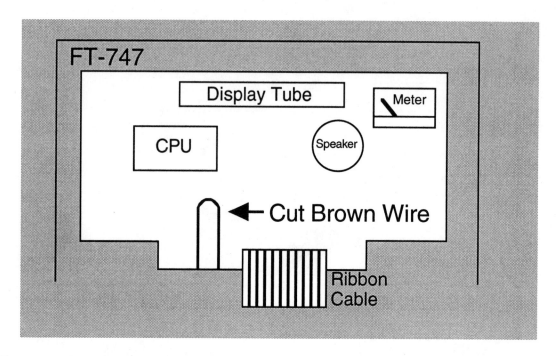

New Range   .5 - 30 MHz

YAESU FT-757GX & FT 757GX II

EXPANDED RF

1. Unplug the DC power cable from the radio.
2. Remove the top cover. You may need to remove the speaker wire to remove the top cover. (see service manual for cover removal)
3. Locate the Black slide switch on the display panel. (to the right of center and halfway down the backside.
4. Use a screwdriver to **set the switch to the left most position.**
5. Reassemble the radio.

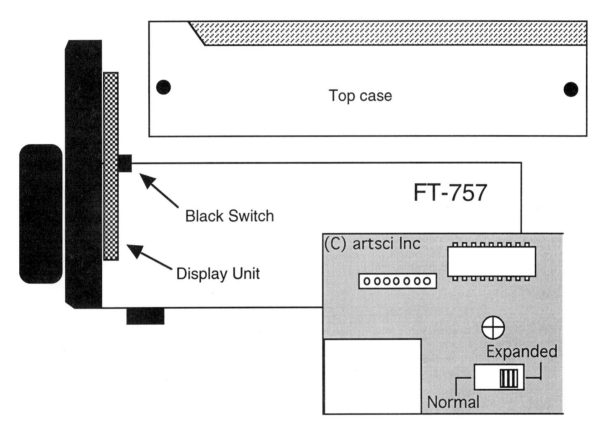

Some models outside the USA may need the following modification -

1. Isolate pin 19 of IC-67(MC68HC05C) on both side of circuit board.
2. Link pin 19 to pin 16 of IC-66(MC14510) with a 10 resistor.
   Be use resistor leads are insulated to brevent shorts.

YAESU FT-767GX

EXPANDED RF

1. Unplug the DC power cable from the radio.
2. Remove any VHF or UHF Band modules.
3. Remove two screws at the front of the top cover and remove the top cover .
4. Locate the GEN/HAM switch inside the shield cover.
5. Use a screwdriver to **set the switch to the GEN position.**
6. Reassemble the radio.

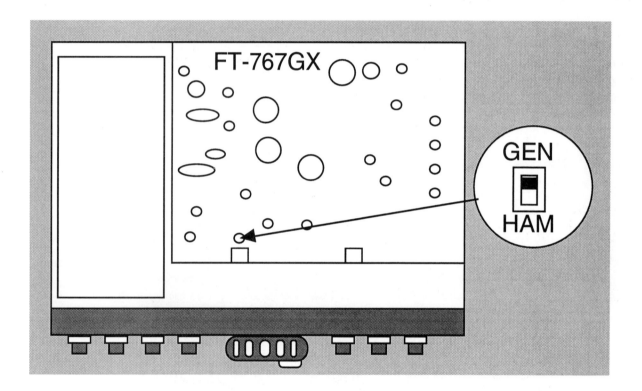

YAESU FT-811

EXPANDED RF & ALIGNMENT CONTROLS
(disables automatic repeater shift)
For Serial # 9D, 9F and 9J series only.
Serial Numbers above 9N can not be modified

1. Remove Battery and Antenna.
2. Remove control knobs, screws,top panel & body screws and open Radio
3. **Remove solder bridge from Pad # 2**
4. **Remove solder bridge from Pad # 4**
4. **Place solder Bridge on Pad # 3**
5. Reassemble the radio
6. **Reset microprocessor.**   (Press and hold [MR] & [VFO] and turn radio on then off)
   (Press and hold both up and down keys and turn power on)
7. Enter the following: 4100 [VFO] 4750 [VFO] 4100 [VFO] 4750 [VFO]
8. Press [Function] & [7] to change channel step.
9. Press [F] & [RPT] and enter offset in both VFO. (5.00 Mhz is standard)

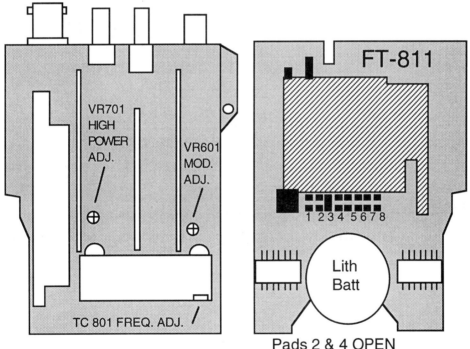

Pads 2 & 4 OPEN
Pad 3 Closed(soldered)

RANGE: RX 410 MHz - 475 MHz
          TX 410 Mhz - 475 MHz

YAESU FT-815
EXPANDED RF
New Range: 410 - 475 MHz RX, 415 - 470 MHz TX
Note: The VCO may need to be adjusted for TX above 460 MHz.

1. Remove Battery and Antenna.
2. Remove the four screws holding the battery track in place.
3. Remove the two black screws holding the rear case in place.
4. Carefully open the front cover from the radio.
5. Locate and **remove the solder from jumper pad #8.**
6. Locate and **solder jumper pads 5 & 7.** Pad 9 is already jumpered.

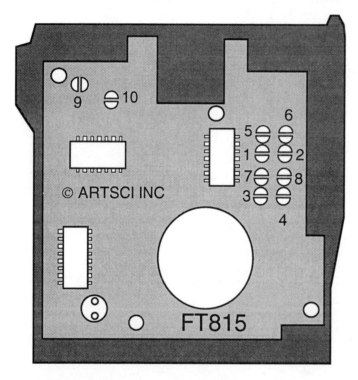

7. Carefully replace the front cover and replace the two black screws.
8. Replace the battery track and the four screws.
9. Reset the microprocessor.
10. **Press and hold [MR], [2] and [VFO] and turn the radio on.**

MORE ---

YAESU FT-815
EXPANDED RF & ALIGNMENT POINTS

11. The radio display will cycle orderly thru the memory channels.
    Enter the following band limits:

    Ch. 1   Enter 410.00 and then press [VFO]   (Rx low limit)
    Ch. 2   Enter 475.00 and then press [VFO]   (Rx high limit)
    Ch. 3   Enter 415.00 and then press [VFO]   (Tx low limit)
    Ch. 4   Enter 470.00 and then press [VFO]   (Tx high limit)

16. Press [F] [0] & [6] and select 5.000 MHz channel spacing in each VFO.

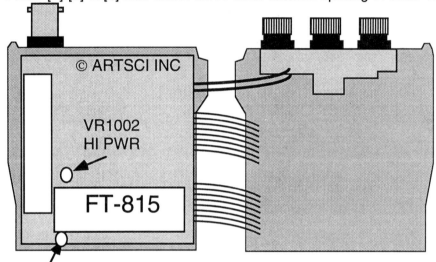

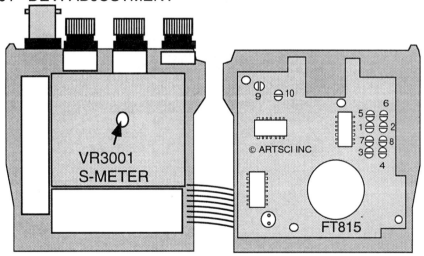

YAESU FT-816
EXPANDED RF
New Range: 400 - 510 MHz RX, 400 - 475 MHz TX
Note: The VCO may need to be adjusted for TX above 460 MHz.

1. Remove Battery and Antenna.
2. Remove the four screws holding the battery track in place.
3. Remove the two black screws holding the rear case in place.
4. Carefully open the front cover from the radio.
5. Locate jumper pads 5 & 7.
6. **Solder jumper pads 5 & 7**. Pad 9 is already jumpered.

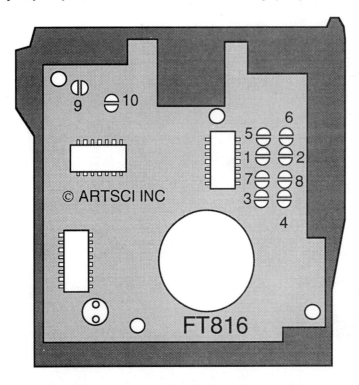

7. Carefully replace the front cover and replace the two black screws.
8. Replace the battery track and the four screws.
9. Reset the microprocessor.
10. **Press and hold [MR], [2] and [VFO] and turn the radio on.**

MORE ---

YAESU FT-816
EXPANDED RF & ALIGNMENT POINTS

11. The radio display will cycle orderly thru the memory channels.
Enter the following band limits:

Ch. 1 Enter 410.00 and then press [VFO] (Rx low limit)
Ch. 2 Enter 510.00 and then press [VFO] (Rx high limit)
Ch. 3 Enter 400.00 and then press [VFO] (Tx low limit)
Ch. 4 Enter 510.00 and then press [VFO] (Tx high limit)

16. Press [F] [0] & [6] and select 5.000 MHz channel spacing in each VFO.

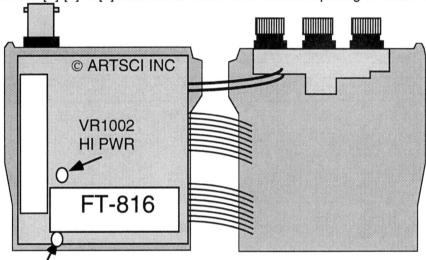

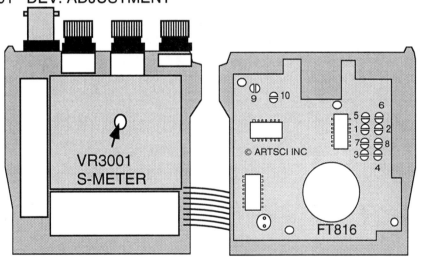

YAESU FT-890

EXPANDED RF (.5 - 30 MHZ)

1. Remove power from the radio.
2. Remove covers.

The next step is done TEMPORARILY.

3. Locate jumper location JW3001 on the DISPLAY UNIT and solder bridge the pads.

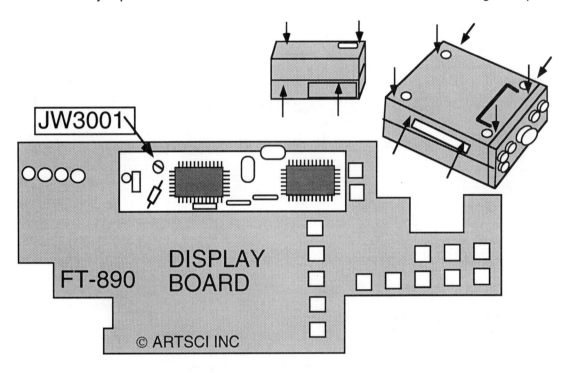

4. Reconnect the power cable.
5. Press and hold [PROC], [AGC-F], [IPO] & [ATT] and turn the power on.
6. Rotate the main dial unitl the display shows 02-ON .
7. Press [PROC]. This will confirm and write the data to EEPROM memory.
8. Turn the power off and remove the power cords.
9. Remove the jumper placed in step 3 above.
10. Replace the covers.

YAESU FT-990
EXPANDED RF

1. Remove power from the radio.
2. Remove the top cover of the transceiver.
3. Locate the Control unit. It is the rightmost of the vertically-mounted circuits boards.
4. Remove the two mounting screws on the boards restraining brackets.
5. Remove the control unit.
6. Locate Jumper pad JP5002. It is located in the next to IC Q5016. IC Q5016 is the rightmost IC of the three large IC in the center of the board.
7. Solder bridge Pad JP5002.
8. Reinstall the Control unit.
9. Locate VR1003 & VR1005 on the RF unit.
10. Connect a 50 Ohm dummy load and a key to the key jack.
11. Set CW mode and the METER to the ALC setting.
12. Dial Frequency 5.000 MHz.
13. Set the RF Power switch fully clockwise.
14. Close PTT and the key. (TRANSMITTING)
15. Adjust VR1003 so that the ALC meter reads to the right edge of the scale.
16. Check frequency range 4.0 - 6.5 MHz to make sure ALC meter reads at least slightly across the entire range.
17. Dial Frequency 8.000 MHz.
18. Adjust VR1005 so that the ALC meter reads to the right edge of the scale.
19. Check frequency range 8.0 - 10.0 MHz to make sure ALC meter reads at least slightly across the entire range.
20. Replace the top cover.

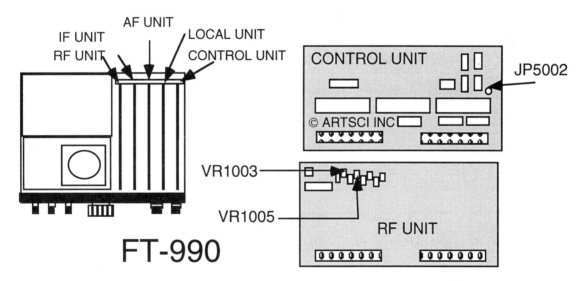

NOTE: Avoid transmissions near 10.940 MHz & 23.60 MHz due to elevated spurious emissions.

YAESU FT-1000

EXPANDED RF

1. Remove power from the radio.
2. Open the case top and bottom.
3. Locate four crews attaching front panel and remove the top screws. Loosen the bottom screws.
4. Tilt front panel forward.
5. On the left side of the radio, remove the plug from the power supply to the front panel. (gray and white wires)
6. Locate jumper position 3 on Control board.
7. **Unsolder the jumper in position 3**
8. Reassemble the radio.
9. Reset the microprocessor.
             (Turn off the Backup Switch, located inside the panel window)

FT-1000 FRONT PANEL

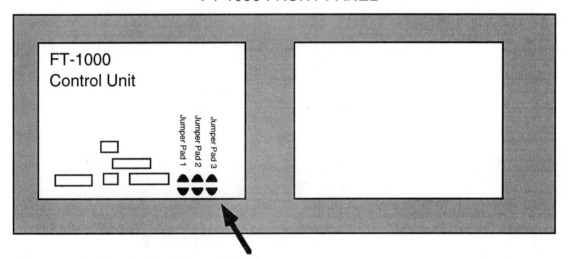

Remove Solder Bridge from Pad #3

New Range: .1 - 30 MHz

YAESU FT-2200

EXPANDED RF
Range 110 - 174 MHz RX
140- 174 MHz TX

1. Remove power and antenna
2. Remove top and bottom covers. (the speaker may fall out)
3. Remove the Volume, Squelch and tuning knobs from the front of the radio.
4. Remove the front panel (push on all four tabs)
5. Remove the tuning knob retainer nut.
6. lift off the LCD display assembly.
7. Locate jumper Pad #5.
8. **Solder jump Pad #5.**
9. Reassemble the radio.
10. Reset the microprocessor.
   (Press and hold [MHz] and [call] buttons and turn the radio on.

Note: A "*" will apear when frequency is below 140 MHz.
   The AM mode will store in memory channels.

Performance Report

Radio _____   Date _____

Owner : Name _____
        Address _____
        City _____ St. _____ Zip _____
        Phone (____) ____ - _____

Description	Before	After
Power out (Low)	_____ Watts	_____ Watts
Power out (High)	_____ Watts	_____ Watts
Frequency Error (Simplex)	_____ Hz	_____ Hz
Frequency Error (Offset)	_____ Hz	_____ Hz
Receive Sensitivity (Mid-band)	_____ uv	_____ uv
Receive Sensitivity (____MHz)	_____ uv	_____ uv
Receive Sensitivity (____MHz)	_____ uv	_____ uv
PL Deviation	_____ Hz	_____ Hz
DTMF Deviation	_____ KHz	_____ KHz
Audio Deviation	_____ KHz	_____ KHz
Lowest usable Freq @ .5 Pwr	_____ MHz	_____ MHz
Highest usable Freq @ .5 Pwr	_____ MHz	_____ MHz

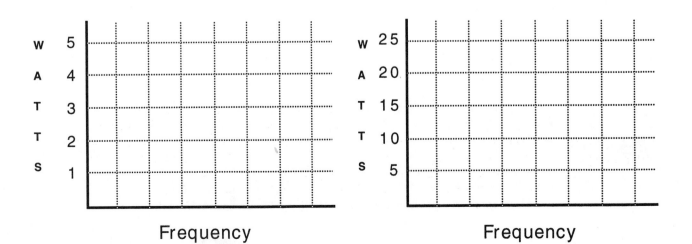

Frequency   Frequency

YAESU FT-2311

EXPANDED RF

1. Remove five screws from the top cover and remove the cover.
2. Remove five screws from the bottom cover and remove the cover.
3. Unplug the speaker.
4. Remove the four screws holding the front panel.
5. Locate jumper pad number 7.
6. **Solder bridge pad number 7.**
7. Locate the reset pins (Located on the front panel and clearly marked).
8. Short the reset pins together for one second.
9. Reassemble the radio.

New range : 1240.00 MHz - 1300.00 MHz

YAESU FT-2400

EXPANDED RF

1. Remove Power and Antenna.
2. Locate and remove the two Allen screws from the front panel.
3. **Locate and unsolder jumper pad 2.**
4. **Locate and solder jump pads 1 & 3.**
5. Reassemble the radio.

New range : 118-174 MHz Rx, 140-174 MHz Tx .

Option #2

1. Follow steps above, except leave solder pad 2 jumpered.
2. Turn radio on and set the upper and lower limits:

Select 138.00 MHz and Press [D/MR] button   (lower RX limit)
Select 174.00 MHz and Press [D/MR] button   (High RX limit)
Select 138.00 MHz and Press [D/MR] button   (lower TX limit)
Select 174.00 MHz and Press [D/MR] button   (High TX limit)

TONE BURST - Solder Pad # 6

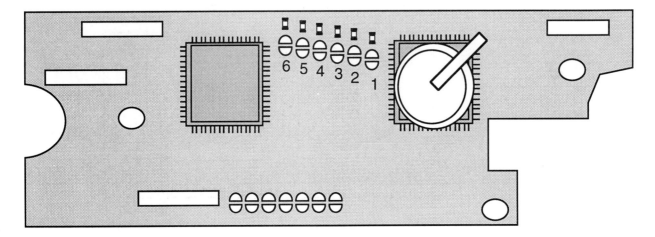

MORE ---

YAESU FT-2400

ALIGNMENT POINTS

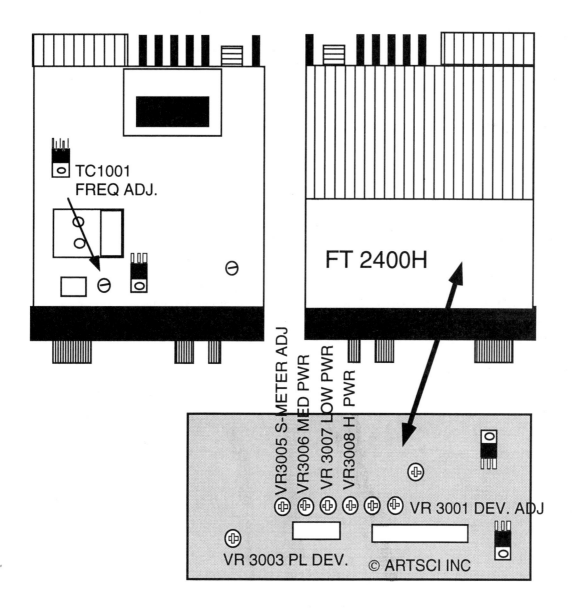

VR3005 S-METER ADJ
VR3006 MED PWR
VR 3007 LOW PWR
VR3008 H PWR

TC1001
FREQ ADJ.

FT 2400H

VR 3001 DEV. ADJ

VR 3003 PL DEV.   © ARTSCI INC

YAESU FT-4700
EXPANDED RF

1. Remove Front Panel.
2. **Locate and jump pads 1,2,5,9,10 & 13.** Solder short them carefully. (The other jumper pads must remain undisturbed)
3. Reassemble radio.
4. Turn power on. (The microprocessor has been reset)
5. Use the [UP] & [DOWN] buttons and dial to set the UHF range as follows :

410.000 MHz	Press [D/MR] button
475.000 MHz	Press [D/MR] button

6. The display will show 47.75 (IF freq. for UHF).   Press [D/MR]
7. Use the up/down buttons and dial to set the VHF range as follows :

138.000 MHz	Press [D/MR] button
174.000 MHz	Press [D/MR] button

8. The display will show 17.3 (IF freq. for VHF).   Press [D/MR]
9. The repeater shifts for both bands are reset to 000. They must be set using the [F] and [PRT] buttons. Refer to page 27 in the user manual.

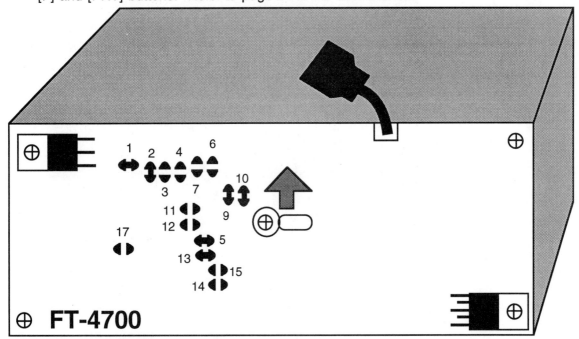

RX Range	138 MHz - 174 MHz	
	410 MHz - 475 MHz	
TX Range	138 MHz - 174 MHz	
	410 MHz - 475 MHz	

MORE ------

YAESU FT-4700
BEEP LEVEL REDUCTION

1. Remove Front Panel
2. Remove the five screws holding Control unit in place.
3. Remove P10 from J04
4. Remove P09 from J03
5. Carefully flip the Control board to access the back side.
6. Locate R08 and R13.
7. Replace R08 and R13 with 560 ohm chip resistors (YAESU # J24205561
8. Reconnect the two Plugs P10 & P09
9. Reassemble the radio.

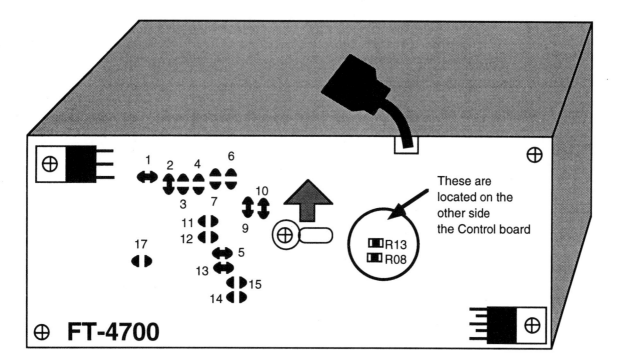

These are located on the other side the Control board

FT-4700

MORE ---

YAESU FT-4700

ALIGNMENT POINTS

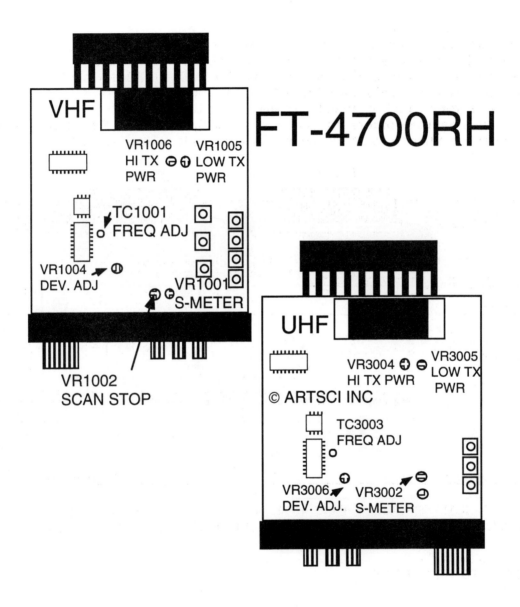

FT-4700RH

VHF

VR1006   VR1005
HI TX ⊖ ⊕ LOW TX
PWR   PWR

TC1001
FREQ ADJ

VR1004 ⊙
DEV. ADJ

VR1001
S-METER

VR1002
SCAN STOP

UHF

VR3004 ⊕ ⊖ VR3005
HI TX PWR   LOW TX
               PWR

© ARTSCI INC

TC3003
FREQ ADJ

VR3006 ⊕   VR3002 ⊖
DEV. ADJ. S-METER ⊕

YAESU FT-5100

EXPANDED RF 128-180 MHZ & 420-475 MHZ
X-BAND Repeater mod & Mic. Band Change mod.

1. Remove power and antenna from the radio.
2. Remove 6 screws from top and bottom covers, remove the covers (watch speaker).
3. Remove the 2 silver screws from each side of the radio securing the control head.
4. Carefully pull the Control Head from the radio. DO NOT REMOVE RIBBON CABLES.
5. **Locate and remove chip resistor R4072.** (RX mod)
6. **Locate and remove chip resistor R4067.** (Mic/Band mod)
7. Locate and **install jumpers in positions R4070, R4068 & R4064.** (RX mod)
NOTE: The circuit board has no numbers: use the picture below to locate chip positions.

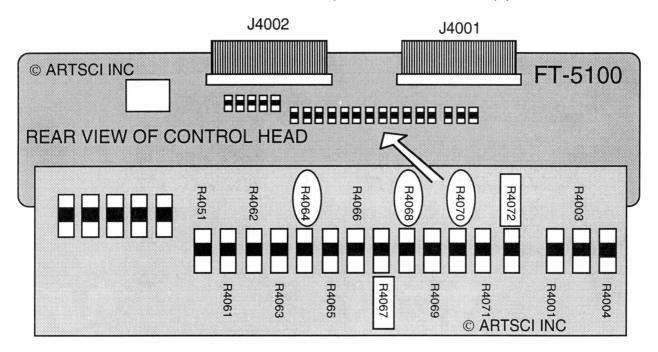

STOCK US JUMPERS: 4001, 4003, 4004, 4051, 4061, 4062, 4067, 4072
POST MOD JUMPERS 4001, 4003, 4004, 4051, 4061, 4062, 4064, 4068, 4070

8. Reassemble the radio.
9. **PROCEED TO NEXT PAGE -**

MORE -

YAESU FT-5100

INITIAL PROGRAMMING INFORMATION
(MUST DO FOR COMPLETE MODIFICATION)

9. Press and hold [D/MR], [F/W] & [REV] and turn power on.
    (The display will show 300.000 & 20.000)
10. Press [MHz] and dial 420.00 and press [D/MR]   - UHF RX low limit
11. Press [MHz] and dial 475.00 and press [D/MR]   - UHF RX high limit
12. Press [MHz] and dial 420.00 and press [D/MR]   - UHF TX low limit
13. Press [MHz] and dial 475.00 and press [D/MR]   - UHF TX high limit
14. Press [MHz] and dial 128.00 and press [D/MR]   - VHF RX low limit
15. Press [MHz] and dial 180.00 and press [D/MR]   - VHF RX high limit
16. Press [MHz] and dial 128.00 and press [D/MR]   - VHF TX low limit
17. Press [MHz] and dial 475.00 and press [D/MR]   - VHF TX high limit
18. Press [F/W] then [RPT] and dial 5.000 and press [RPT] - UHF offset
19. Press [F/W] then [REV] and dial 25.0 and press [RPT].
20. Press [BAND] then [F/W] then [RPT] and dial 0.600 and press [RPT] - VHF offset.

* * * *

SOFT RESET (Memory clear) - Press and hold [D/MR] & [REV] and turn radio on.

CROSS-BAND REPEATER OPERATION

1. Select the desired VHF & UHF frequencies
2. Select low power transmit on both bands (To protect your radio)
3. If desired, adjust the TX time out timer value. (The default is 15 minutes)
    To adjust: Press and hold [LOW] & turn power on.
                   Dial desired time out value (0-60 minutes)
                   Turn radio off.

TURN ON -   Press and hold [RPT] and turn radio on.
TURN OFF -   Press and hold [RPT] and turn radio on.

EXTRA Modification

Remove solder from Jumper R4067 to make Microphone [D/MR] button switch band on the radio.

YAESU FT-5100
ALIGNMENT POINTS

VR1003 UHF
TX OUTPUT

VR1002
UHF AFP

VR1001
UHF AFP

© artsci

VR1006 VHF
DEVIATION

VR1007 UHF
DEVIATION

FT-5100

VHF AFP TEST POINT

VR404 - UHF SCANNER CENTER-STOP

VR406 - UHF-S-METER CAL

VR402 - UHF SQUELCH PRESET

VR101-
VHF AFP

VR103-
VHF AFP

VR401-VHF SQUELCH PRESET

TP403 - UHF SCANNER
CENTER-STOP TEST

VR405 - VHF S-METER CAL

VR402 - VHF SCANNER CENTER-STOP

TP401/TP402 - VHF SCANNER
CENTER-STOP TEST

YAESU FT-5200

EXPANDED RF

1.   Remove power from the radio.
2.   Release and remove the front panel.
3.   Remove the six screws from the top cover of the radio.
4.   Remove the six screws from the bottom of the radio.
5.   Remove the top and bottom covers.
     (CAUTION: the speaker might fall out.)
6.   Remove the two screws & front control head mounting plate from the radio.
7.   Locate solder pads 1 - 7.
     (Standard jumpered pads are 2 and 7 only)
8.   **Solder jump pads 1,3 and 6**
     (Pads 1,2,3,6 & 7 are now jumpered)
9.   **Unsolder jump pad 17.** (X-Band repeater mod) May be done at the factory!
     Caution: Be sure to work on PAD 17. see drawing below

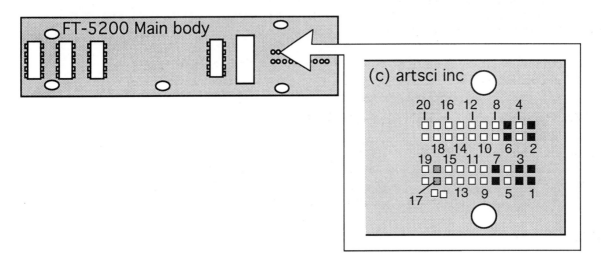

See Next page for further instructions.

MORE ---

YAESU FT-5200

EXPANDED RF

10. Install front panel mounting plate.
11. Reassemble the radio.
12. Reconnect the power to the radio.
13. **Press and hold [D/MR], [F/W] & [REV] keys and turn radio on.**
   (Display will show 000.000 & 300.000 on the display)
14. **Set the VHF Receive and Transmit limits:**

   Enter 118.00 MHz and press [D/MR]   (VHF RX Low)
   Enter 174.00 MHz and press [D/MR]   (VHF RX High)
   Enter 140.00 MHz and press [D/MR]   (VHF TX Low)
   Enter 174.00 MHz and press [D/MR]   (VHF TX High)

15 **Set the UHF Receive and Transmit limits:**

   Enter 420.00 MHz and press [D/MR]   (UHF RX Low)
   Enter 475.00 MHz and press [D/MR]   (UHF RX High)
   Enter 420.00 MHz and press [D/MR]   (UHF TX Low)
   Enter 475.00 MHz and press [D/MR]   (UHF TX High)

16. Press [Function] then [REP] and select 5 MHz Repeater offset for UHF band.
17. Press [Function] then [REP] and select 600 kHz Repeater offset for UHF band.

To activate X-Band repeater function:

   Press and hold [RPT] and turn power on.
      It is recommended that you unplug the microphone during X-Band
   operation. (The Mic is live)
      • Adjust the volume control to adjust repeat audio level.

Options:

Override automatic display dimmer:
   Press and hold [MHz] and turn radio on: Use Channel knob to select brightness.

Keyboard VHF Expanded Receive:
   Press and hold [DVS] & [MHz] keys and turn radio on.

MORE ---

YAESU FT-5200

ALIGNMENT POINTS

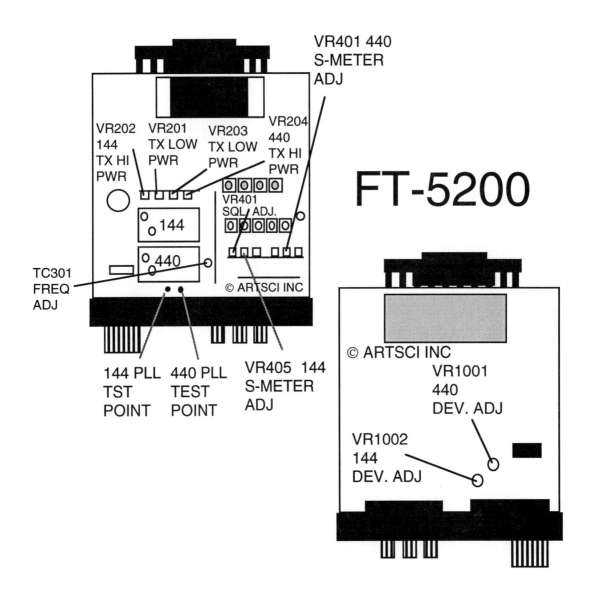

Performance Report

Radio _____   Date _____

Owner : Name _____
        Address _____
        City _____ St. ____ Zip _____
        Phone (____) ____ - _____

Description	Before	After
Power out (Low)	_____ Watts	_____ Watts
Power out (High)	_____ Watts	_____ Watts
Frequency Error (Simplex)	_____ Hz	_____ Hz
Frequency Error (Offset)	_____ Hz	_____ Hz
Receive Sensitivity (Mid-band)	_____ uv	_____ uv
Receive Sensitivity (____MHz)	_____ uv	_____ uv
Receive Sensitivity (____MHz)	_____ uv	_____ uv
PL Deviation	_____ Hz	_____ Hz
DTMF Deviation	_____ KHz	_____ KHz
Audio Deviation	_____ KHz	_____ KHz
Lowest usable Freq @ .5 Pwr	_____ MHz	_____ MHz
Highest usable Freq @ .5 Pwr	_____ MHz	_____ MHz

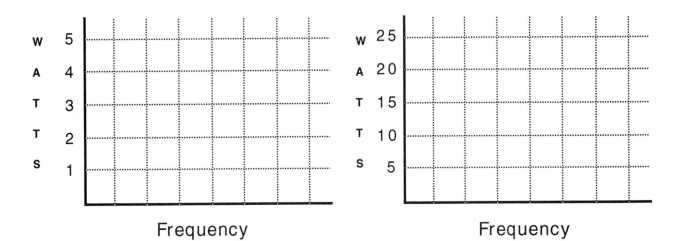

W A T T S   5 4 3 2 1

Frequency

W A T T S   25 20 15 10 5

Frequency

YAESU FT-6200
EXPANDED RF (420 - 475 MHz) / X-Band repeater

1. Remove power from the radio.
2. Release and remove the Control head.
3. Remove the top and bottom covers. Six screws hold the top and bottom covers on.
4. Remove the two silver screws holding the control head mounting bracket.
5. Remove the mounting bracket.
6. **Locate and solder jumper pad #6.**
   Pads 2, 4, 6, 7, 8, 15, 17 & 18 will now be jumpered.
7. Locate and **remove solder jumper pad #17.** (X-Band repeater mod)
   Caution: Make sure you jumper the proper pad. see drawing below.
8. Reassemble the radio.
9. Reconnect the power.
10. **Press and hold [D/MR], [F/W] & [REV] and turn the power on.**
    The radio will now show 300.000
11. **Enter the following band limits:**

    420.00 and then press [D/MR] (UHF Rx low limit)
    475.00 and then press [D/MR] (UHF Rx high limit)
    420.00 and then press [D/MR] (UHF Tx low limit)
    475.00 and then press [D/MR] (UHF Tx high limit)
12. Press [FUNCTION] and then [RPT] and select 5.000 MHz repeater offset.

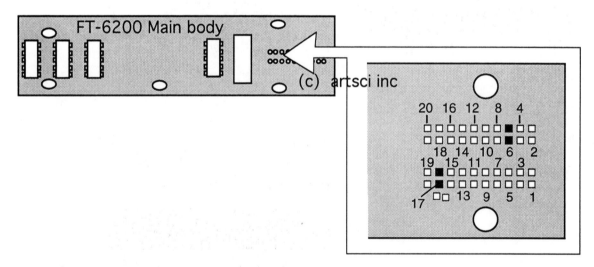

To activate X-Band repeater function: Press and hold [RPT] and turn power on.
To override automatic display dimmer: Press and hold [MR] and turn power on and select the desired brightness level)

MORE ---

YAESU FT-6200

ALIGNMENT CONTROLS

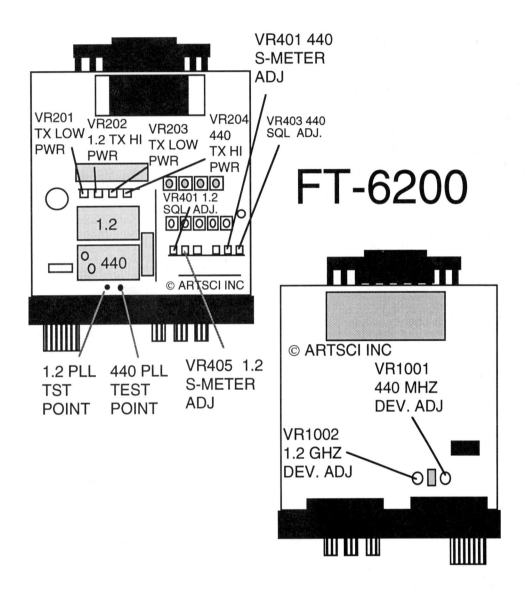

VR401 440
S-METER
ADJ

VR201
TX LOW
PWR

VR202
1.2 TX HI
PWR

VR203
TX LOW
PWR

VR204
440
TX HI
PWR

VR403 440
SQL ADJ.

VR401 1.2
SQL ADJ.

1.2

440

© ARTSCI INC

FT-6200

1.2 PLL
TST
POINT

440 PLL
TEST
POINT

VR405 1.2
S-METER
ADJ

© ARTSCI INC

VR1001
440 MHZ
DEV. ADJ

VR1002
1.2 GHZ
DEV. ADJ

YAESU FT-7400

EXPANDED RF (420 - 470 MHz)

1. Remove power from the radio.
2. Remove Front Panel.
3. Locate solder pad #1. (Behind front control panel)
4. **Solder jump pad # 1**
5. Reassemble the radio.

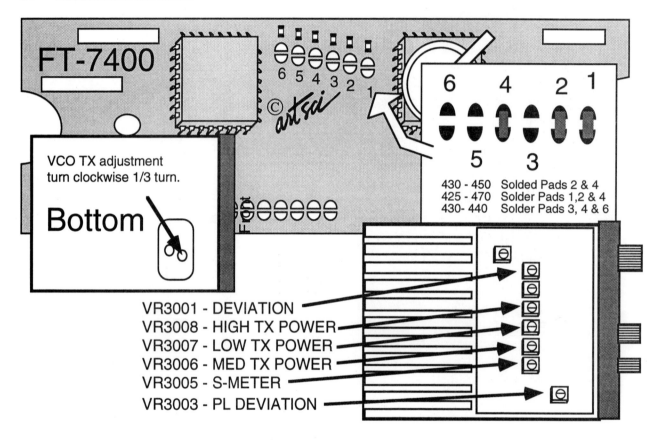

FT-7400

6 5 4 3 2 1

© artsci

6		4		2	1

5   3

430 - 450   Solded Pads 2 & 4
425 - 470   Solder Pads 1,2 & 4
430- 440   Solder Pads 3, 4 & 6

VCO TX adjustment
turn clockwise 1/3 turn.

Bottom

Front

VR3001 - DEVIATION
VR3008 - HIGH TX POWER
VR3007 - LOW TX POWER
VR3006 - MED TX POWER
VR3005 - S-METER
VR3003 - PL DEVIATION

YAESU FT-2070
EXPANDED RF

1. Remove battery and Antenna from the radio.
2. Remove screws and open case
3. Locate and **unsolder jumper pad as shown below**
   (Pad connected to Microprocessor pin 11)
4. Reassemble the radio.
5. Reset the Microprocessor
   (Press [PRI] and turn the radio on.)

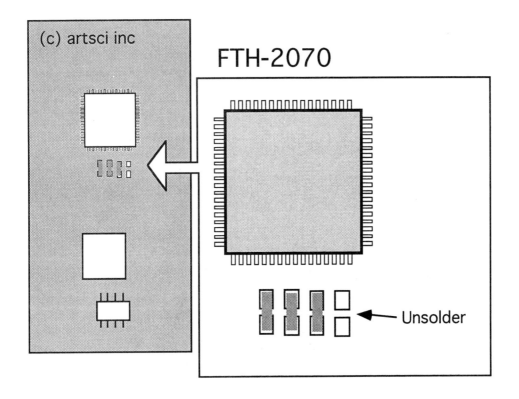

New Range: 134 - 174 MHz & 400 - 500 MHz

YAESU FT-ONE

EXPANDED RF

1. Unplug the power from the radio.
2. Open radio and locate the CONTROL UNIT.
3. Locate and **install a Jumper between Point A and point B.** No Jumper to point C.
4. **Remove any jumper to point D.** (Transmit range point)
5. Reassemble radio.

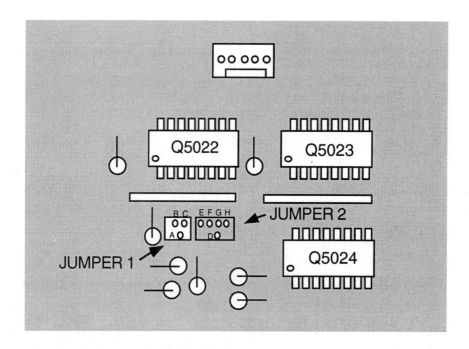

RX Range   150 KHz - 30 MHz
TX Range   1.8 MHz - 30 MHz

YAESU NC-29

TRICKLE MODE

This modification will allow you to select the amount of time used to fast charge your battery pack. The standard NC-29 will fast charge a battery for five hours and then switch to trickle charge every time a battery is inserted, even if the battery is fully charged.

This modification will provide a push button to speed up the Internal clock. By pressing the button, you can watch the time remaining LEDs on the panel and select the amount of full charging time.

1. Unplug the charger for the AC power
2. Locate IC Q02. see drawing
3. Solder tack a 390 Ohm 1/2 watt resistor and a normally open push button to Pins 13 & 15
4. Position the push button switch in a handy position on the plastic case.

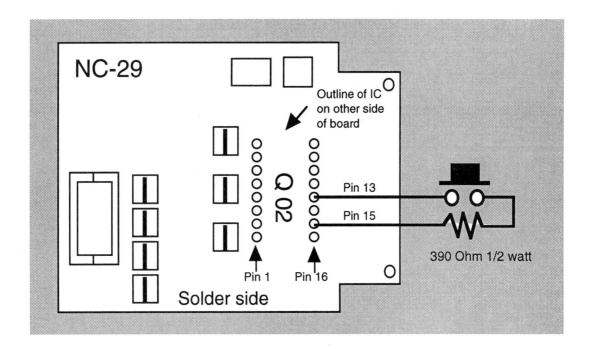

YAESU NC-42

Charging additional batteries

This modification will allow you to charge FNB-12S, FNB-14, FNB-17, FNB-25, FNB-26 and FNB-27 batteries.

1. Remove the ridge on the inside of the battery charging cup. (right side only)

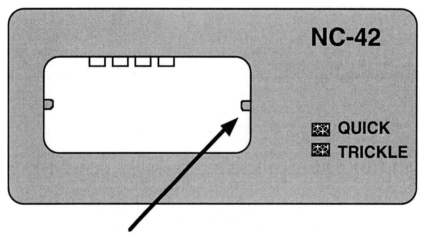

Remove this ridge. Use a file or similar tool

Charging time for all batteries should be about 1 hour or less.

YAESU Hand Held to Packet TNC

FT-23,33,73,109,209,709,727,470,411,811,911

Parts required:

    1 - 0.1 uF, 50V Disk Ceramic Cap
    2 - 2.2k Ohms, 1/4 Watt Resistor
    1 - 2.5 mm audio plug
    1 - 3.5 mm audio plug

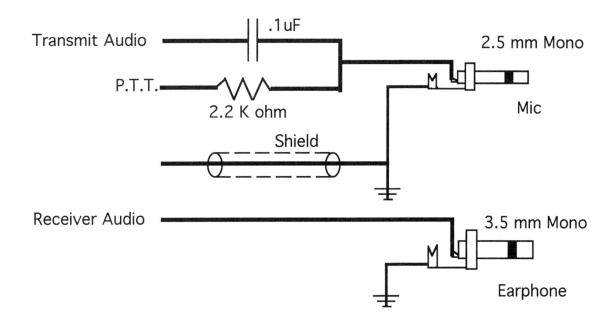

Yaesu Reset Commands

Radio	Function	Command
FT-1000	Hard Reset	Flip off BACKUP switch. (Inside the top panel window)
	Memory Reset	Press & hold [SUB] & [ENTER] & turn power on
	Soft Reset	Press & hold [1.5] & [3.5] & turn power on. (For checking Display and ROM version)
FT-990	Hard Reset	Flip off BACKUP switch. (Inside the top panel window)
	Memory Reset	Press & hold [GEN] & [ENT] & turn power on
	Soft Reset	Press & hold [1.5] & [3.5] & turn power on. (For checking Display and ROM version)
FT-890	Hard Reset	Press & hold [HAM/GEN] & [CLAR] & turn power on.
	Soft Reset	Press & Hold [A/B] & [A=B] & turn power on (For checking Display and ROM version)
FT-767GX	Hard Reset	Switch [B.U.] off & turn radio on.
	Freq. Range Reset	Press and hold [OFFSET] & turn power on. (140.00 - 148.99 MHz) Press and hold [CLAR] & turn power on. (140.00 - 145.99 MHz) Press and hold [MCK] & turn power on. (140.00 - 1487.99 MHz)
	430/440 toggle	Press and hold [0] & turn power on.
FT-757GX	Hard Reset	Press & hold [MARKER] & [LINEAR] & turn power on.
FT-747GX	Hard Reset	Slide Backup switch towards tuning dial. (Located on bottom of panel)
FRG-8800	Hard Reset	Remove backup batteries
FRG-100	Hard Reset	Turn off backup switch on rear of radio for 5 seconds.

Yaesu Reset Commands

Radio	Function	Command
FT-26		
FT-76	Ham/Extended RX	Press and hold [UP] & [DOWN] & turn on.
	Factory Defaults Soft Reset (memory clear)	Press and hold [T] & [REV] & turn on.
	Master Reset	Press and hold [D/MR] & [T] & [REV] & turn on. (must enter new band limits)
FT-411E		
FT-811		
FT-911		
FT-415		
FT-470		
FT-815		
FT-530	Ham/Extended RX	Press and hold [UP] & [DOWN] & turn on.
	Factory Defaults	Press and hold [T] & [REV] & turn on.
FT-2400H	Ham/Extended RX	Press and hold [UP] & [DOWN] & turn on
	Memory Reset	Press [D/MR] & [F/w] & turn on.
	Factory Defaults	Press [D/MR] & [REV/SKIP] & turn on & turn off & Press & hold [D/MR] & turn on.
FT-5100	Factory Defaults	Press and hold [D/MR] & [REV] & turn on.
FT-5200	Ham/Extended RX	Press and hold [MHz] & [DVS/HOLD] & turn on.
	Factory Defaults	Press and hold [D/MR] & [REV] & turn power on.
FT-212		
FT-712		
FT-912	Ham/Extended RX	Press and hold [MHz] & [VOICE] & turn power on.

Yaesu Reset Commands

Radio	Function	Command
FT-290 FT-690 FT-790II	Hard Reset	Switch internal backup switch off of 30 seconds.
FT-736R	Hard Reset	Switch internal backup switch off of 30 seconds.

Radio / Tech Modifications

OTHER MANUFACTURES

Performance Report

Radio _____   Date _____

Owner : Name _____
           Address _____
           City _____ St. _____ Zip _____
           Phone (____) _____ - _____

Description	Before	After
Power out (Low)	_____ Watts	_____ Watts
Power out (High)	_____ Watts	_____ Watts
Frequency Error (Simplex)	_____ Hz	_____ Hz
Frequency Error (Offset)	_____ Hz	_____ Hz
Receive Sensitivity (Mid-band)	_____ uv	_____ uv
Receive Sensitivity (____MHz)	_____ uv	_____ uv
Receive Sensitivity (____MHz)	_____ uv	_____ uv
PL Deviation	_____ Hz	_____ Hz
DTMF Deviation	_____ KHz	_____ KHz
Audio Deviation	_____ KHz	_____ KHz
Lowest usable Freq @ .5 Pwr	_____ MHz	_____ MHz
Highest usable Freq @ .5 Pwr	_____ MHz	_____ MHz

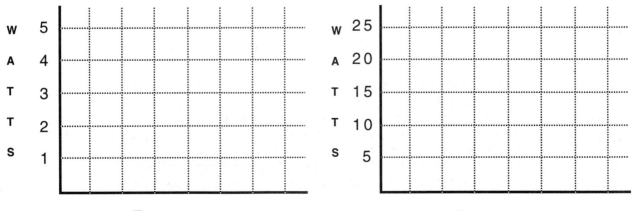

Frequency   Frequency

AZDEN   PCS-6000

EXPANDED RF

1. Remove Power and Antenna.
2. Remove the Top and Bottom covers.
3. Locate and remove the four flat Phillips screws that secure the display to the chassis.
4. Locate and remove the four small Phillips screws securing the PC Board to the chassis.
5. Locate and remove the one Phillips screw above the Microphone connector.
6. Carefully remove the PC board. CAUTION: Do not bend the PIN connectors.
7. Locate and **remove Diode D-207.** (Unsolder or Cut the diode away)
8. Reassemble the radio.

RANGE: 138.000 MHz - 160.000 MHz

AZDEN   PCS-7000

EXPANDED RF

1. Remove Power and Antenna.
2. Remove the Top and Bottom covers.
3. Locate and remove the four flat Phillips screws that secure the display to the chassis.
4. Locate and remove the four small Phillips screws securing the PC Board to the chassis.
5. Locate and remove the one Phillips screw above the Microphone connector.
6. Carefully remove the PC board. CAUTION: Do not bend the PIN connectors.
7. Locate and **remove Diode D-207.** (Unsolder or Cut the diode away)
8. Reassemble the radio.

RANGE: 138.000 MHz - 160.000 MHz

AZDEN   AZ-21A

EXPANDED RF

1.   Remove Power and Antenna.
2.   Remove Speaker & Squelch knobs
3.   Remove battery rail screws
4.   Remove three back cover screws.
5.   Remove top cover and rubber gasket
6.   Separate radio. (open like a book)
7.   Remove three screws from right hand board and move aside
8.   Locate lower board and solder pads B0 through B5
9.   Locate and **solder bridge pads B0 & B1.**
10. Reassemble the radio.
11. **Reset the microprocessor**
     (Hold down the [CLR] key and turn the radio on)

KDK   FM-240

EXPANDED RF

1. Remove Power and Antenna.
2. Remove the cover.
3. **Press the RESET Button**.
4. **Enter the new limits on the front panel switch.**   (Range 140-156 MHz)
8. Reassemble the radio.

RANGE: 140.00 MHz - 156.00 MHz

KDK   FM-2033

EXPANDED RF

1.   Remove Power and Antenna.
2.   Remove screws and open the case.
3.   **Connect diode D-21 (ECG-519) to Module INT-2033.**
4.   Reassemble the radio.

TEN TEC   PARAGON

EXPANDED RF

1. Remove Power and Antenna.
2. Remove the Top cover.
3. Locate and **clip small jumper labeled "HAM"**.
8. Reassemble the radio.

RANGE: 1.7 MHz - 30 MHz

Performance Report

Radio _____   Date _____

Owner : Name _____
       Address _____
       City _____ St. ____ Zip _____
       Phone (____) ____ - _____

Description	Before		After	
Power out (Low)	_____	Watts	_____	Watts
Power out (High)	_____	Watts	_____	Watts
Frequency Error (Simplex)	_____	Hz	_____	Hz
Frequency Error (Offset)	_____	Hz	_____	Hz
Receive Sensitivity (Mid-band)	_____	uv	_____	uv
Receive Sensitivity (____MHz)	_____	uv	_____	uv
Receive Sensitivity (____MHz)	_____	uv	_____	uv
PL Deviation	_____	Hz	_____	Hz
DTMF Deviation	_____	KHz	_____	KHz
Audio Deviation	_____	KHz	_____	KHz
Lowest usable Freq @ .5 Pwr	_____	MHz	_____	MHz
Highest usable Freq @ .5 Pwr	_____	MHz	_____	MHz

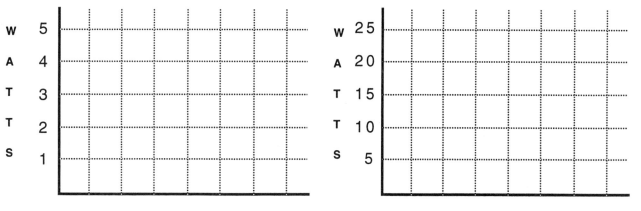

          Frequency   Frequency

RANGER AR-3300

EXPANDED RF

1.   Turn radio on and enter the following:

[ENTER]   [1 CH]   [ENTER]   [MANUAL]   [ENTER] [100 HZ DOWN]

[ENTER] [MEMORY] [MANUAL] [SCAN] [PROGRAM]

[100 HZ UP] [ENTER] [ENTER]

PUSH   [1 MHZ UP] UNTIL 29.933.0   APPEARS

[ENTER] [SCAN DOWN] [ENTER] [2 CH] [ENTER]

[SCAN DOWN]

OPEN THE SQUELCH

The radio will now scan down in 10kHz steps. Store
desired Frequencies into memory channels for later use.

OR

Solder jump the 3 pins located on the back side of the circuit board near the front center.

More ---

RANGER AR-3300

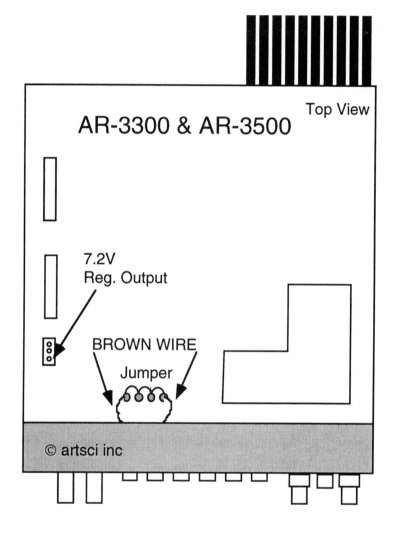

Top View

AR-3300 & AR-3500

7.2V
Reg. Output

BROWN WIRE

Jumper

© artsci inc

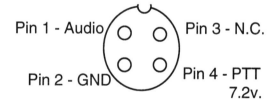

Pin 1 - Audio   Pin 3 - N.C.

Pin 2 - GND   Pin 4 - PTT
7.2v.

RANGER AR-3500

EXPANDED RF

1.   Turn radio on and enter the following:

[ENTER]   [1 CH]   [ENTER]   [MANUAL]   [ENTER] [100 HZ DOWN]

[ENTER] [MEMORY] [MANUAL] [SCAN] [PROGRAM]

[100 HZ UP] [ENTER] [ENTER]

PUSH   [1 MHZ UP] UNTIL 29.933.0   APPEARS

[ENTER] [SCAN DOWN] [ENTER] [2 CH] [ENTER]

[SCAN DOWN]

OPEN THE SQUELCH

The radio will now scan down in 10kHz steps. Store
desired Frequencies into memory channels for later use.

OR

Solder jump the 3 pins located on the back side of the circuit board near the front center.

More ---

RANGER AR-3500

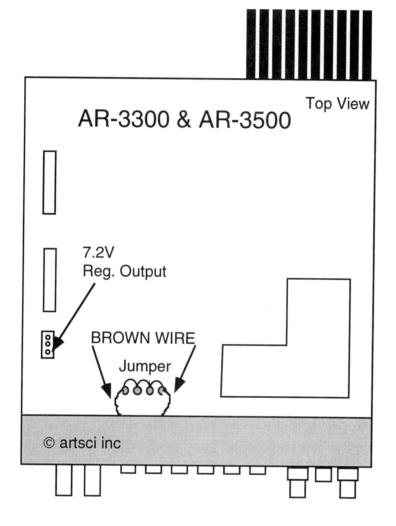

AR-3300 & AR-3500

Top View

7.2V
Reg. Output

BROWN WIRE

Jumper

© artsci inc

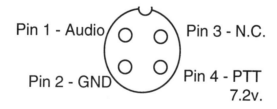

Pin 1 - Audio
Pin 3 - N.C.
Pin 2 - GND
Pin 4 - PTT
7.2v.

Performance Report

Radio _____   Date _____

Owner : Name _____
        Address _____
        City _____ St. ____ Zip _____
        Phone (____) ____-_____

Description	Before		After	
Power out (Low)	_____	Watts	_____	Watts
Power out (High)	_____	Watts	_____	Watts
Frequency Error (Simplex)	_____	Hz	_____	Hz
Frequency Error (Offset)	_____	Hz	_____	Hz
Receive Sensitivity (Mid-band)	_____	uv	_____	uv
Receive Sensitivity (____MHz)	_____	uv	_____	uv
Receive Sensitivity (____MHz)	_____	uv	_____	uv
PL Deviation	_____	Hz	_____	Hz
DTMF Deviation	_____	KHz	_____	KHz
Audio Deviation	_____	KHz	_____	KHz
Lowest usable Freq @ .5 Pwr	_____	MHz	_____	MHz
Highest usable Freq @ .5 Pwr	_____	MHz	_____	MHz

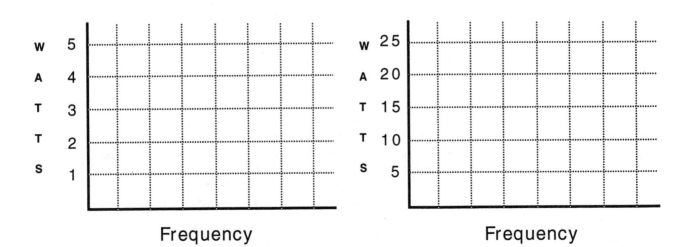

W A T T S   5 4 3 2 1   Frequency

W A T T S   25 20 15 10 5   Frequency

UNIDEN HR-2500
EXPANDED RF

1. Remove Power and Antenna.
2. Remove screws and open the case.
3. Locate synthesizer board on the bottom of the radio.
4. If your radio has microprocessor # UC-1208
   Unsolder and lift pins 28 & 29 of the microprocessor.
   > You may wish to leave the pin soldered and etch the ground trace
   > Go to instruction #6

5. If your radio's microprocessor is NOT a UC-1208
   Unsolder and lift pins 20 & 21 of the microprocessor.
   > You may wish to leave the pin soldered and etch the ground trace
   > Go to instruction #6

6. Connect the lifted pins together and jumper these pins to +5 volts with a 10K resistor
   > +5 volts can be found on the 7805 voltage regulator
   > or
   > from the Cap. right next to pins 28 & 29.
6. Reassemble the radio.

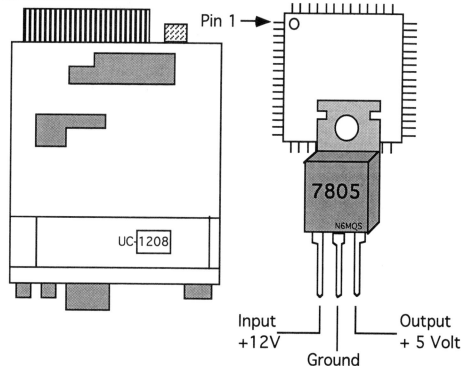

UNIDEN HR-2510

EXPANDED RF

1. Remove Power and Antenna.
2. Remove screws and open the case.
3. Locate the Synthesizer board.
4. Pins 34 & 35 are grounded together on the underside of the synthesizer board. Cut the traces that connect these two pins to ground. (Cut all traces to these pins)
5. Solder one side of a 10K resistor to the connecting point of pins 34 & 35.
6. Connect the other leg of the 10 K resistor to + 5 volts. (where R181 & 187 are connected together.
7. Reassemble radio

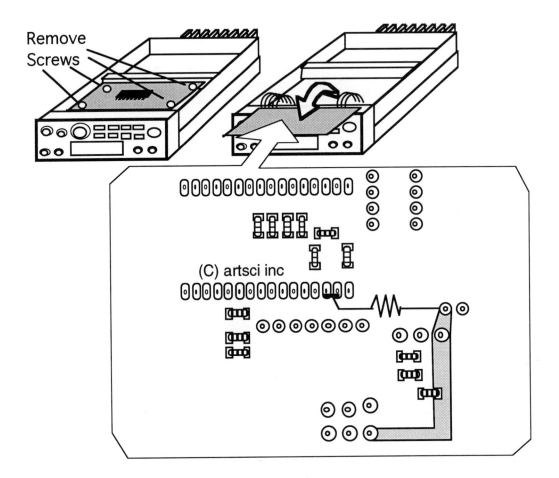

COVERAGE : 26.0000 to 29.9999 MHz

UNIDEN HR-2510

ALIGNMENT POINTS

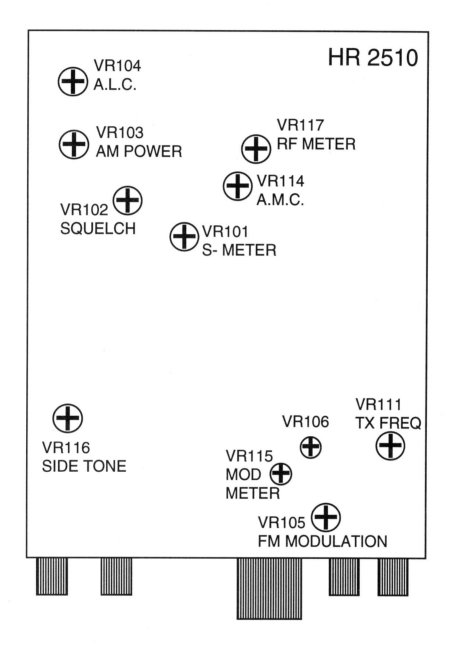

HR 2510

VR104 A.L.C.

VR103 AM POWER

VR117 RF METER

VR114 A.M.C.

VR102 SQUELCH

VR101 S- METER

VR116 SIDE TONE

VR106

VR111 TX FREQ

VR115 MOD METER

VR105 FM MODULATION

UNIDEN HR-2600
EXPANDED RF

You will need to replace the microprocessor. Replacement part # is UC-1250. You will lose the repeater offset.

1.   Remove Power and Antenna.
2.   Remove screws and open the case.
3.   Locate the Synthesizer board.
4.   Pins 34 & 35 are grounded together on the underside of the synthesizer board. Cut the traces that connect these two pins to ground.
5.   Solder one side of a 10K resistor to the connecting point of pins 34 & 35.
6.   Connect the other leg of the 10 K resistor to + 5 volts. (where R181 & 187 are connected together.
7.   Reassemble radio

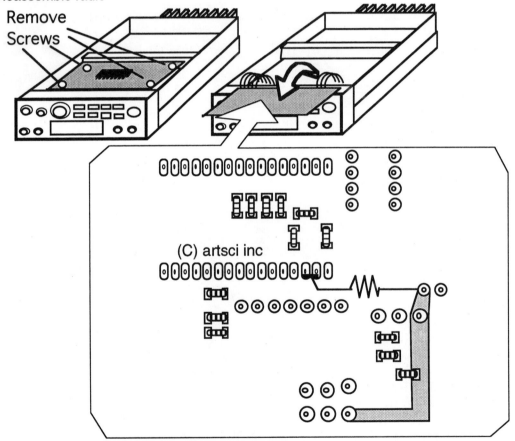

COVERAGE : 26.0000 to 29.9999 MHz

Radio Shack HTX-100
EXPANDED RF

1. Remove Power and Antenna.
2. Remove screws and open the case.
3. Locate synthesizer board on the bottom of the radio.
4. If your radio has microprocessor # UC-1208
    Unsolder and lift pins 28 & 29 of the microprocessor.
    You may wish to leave the pin soldered and etch the ground trace
    Go to instruction #6

5. If your radio's microprocessor is NOT a UC-1208
    Unsolder and lift pins 20 & 21 of the microprocessor.
    You may wish to leave the pin soldered and etch the ground trace
    Go to instruction #6

6. Connect the lifted pins together and jumper these pins to +5 volts through a 10K resistor
    (+5 volts can be found on the 7805 voltage regulator
    or
    from the Cap. right next to pins 28 & 29.)
6. Reassemble the radio.

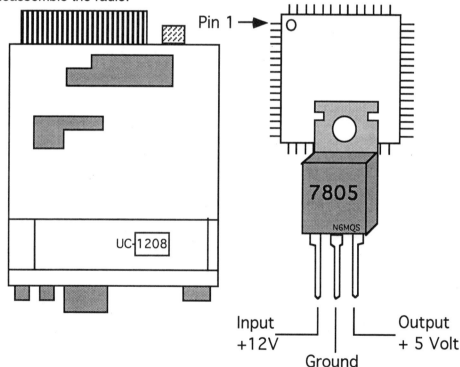

Pin 1 →

UC-1208

7805

N6MQS

Input ___ Output
+12V ___ + 5 Volt
Ground

RCI 2950

Clarifier Fine Tune (Tracks both TX & RX)
Expanded Range
CB "Style" operation
Instant Channel 9

1. Remove Power and Antenna.
2. Remove screws and open the case.
3. **Remove Diode D59.**
4. **Cut lead on Resistor R197.**(see Drawing)
5. **Apply +8 volts from regulator to Resistor R 197**. (see Drawing)
6. Reassemble the radio.

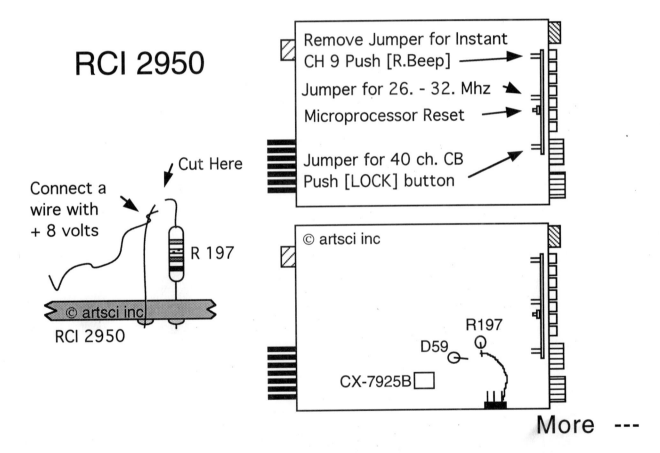

RCI 2950

Cut Here

Connect a
wire with
+ 8 volts

R 197

© artsci inc

RCI 2950

Remove Jumper for Instant
CH 9 Push [R.Beep]
Jumper for 26. - 32. Mhz
Microprocessor Reset

Jumper for 40 ch. CB
Push [LOCK] button

© artsci inc

R197
D59
CX-7925B

More ---

RCI 2950

Alignment Procedure

1. Set the frequency to 26.000 MHz (any mode)

2. Connect a DC voltmeter between J13 and ground.
    (The chassis is not grounded. You can find ground on the main circuit board)
    Adjust L17 to obtain a 1.0 V reading.

3. Set the service monitor to 10.240 MHz, SSB mode.
    Sniff at X2 and zero beat using VC2.

4. Remove the shorting bar located near the final amplifier transistors and key the radio.
    Sniff X2 and adjust VR21 to zero beat.

5. Repeat step 4 for receive at X1.

6. Set the service monitor to 10.695 MHz.
    Key the transmitter and sniff X3 in either AM or FM.
    Adjust L27 and zero beat.

7. Un-key the radio.
    Set the service monitor to 10.6925 MHz, USB mode.
    Key the transmitter and adjust L29 to zero beat.
    Un-key.

8. Un-key the radio.
    Set the service monitor to 10.6975 MHz, LSB mode.
    Key the transmitter and adjust L28 to zero beat.
    Un-key.

More ---

RCI 2950

Alignment Procedure Part 2

9.   Replace shorting bar and set the radio to 28.0500 MHZ FM mode.

10.   Inject an on-frequency FM signal into the radio and tune for best SINAD
       by adjusting L8, L9, L11, L12, L14, L4, L3, L5, L6 and L7.
       Repeat this step until SINAD reading of 12db or better with a .2 uV
       input.

11.   Key the radio in UBS with a 1 KHz tone at 30 mV at the mic input.
       Adjust VR12 for maximum, approximately 30 W.

12.   Adjust VC3, L34, L43, L46, L47, L48 and L19 for peak power out.
       Adjust VR12 to set max power to 25 watts.

13.   Set mode to FM and key the radio.
       Set the output power to 10 watts using VR13.

14.   Set the mode to AM and adjust VR14 for 90% modulation.

***** radio is now aligned.

HEATHKIT SB-1400

EXPANDED RF

1.	Turn the radio on.
2.	Set display to 12.3456
3.	Press [BAND] button.
4.	Turn radio off.
5.	Turn radio on.

Note:	You must perform these steps within 3 seconds to properly reset the radio.

Sender TR-450

EXPANDED RF
400 - 469.996 MHz

1.   Press [F] and turn power on, then off (RESET Radio)
2.   Press [3] and turn power on. (400 - 469 MHz RX)
3.   Press [F] &[0] then set CTCSS to 88.5 MHz (use rotary knob)
4.   Press [F] & # then set page code to (memory 0 = C000)
5.   Press [F] & [3] then set channel step to 5 KHz
6.   Press [F] & [9] then keyin 6.1 MHz
7.   Press [F[& [0] then [8]
8.   Press the [*/ENT] key

Note: during testing, these steps needed to be performed multiple times

Radio / Tech Modifications

CB Modifications

C.B. RADIOS

Radio / Tech Modifications

CB Modifications

COBRA 148GTL
any other CB using MB8719 IC

EXPANDED RF

Note: This mod requires seven toggle switches to control Frequency. See frequency
chart on the next page.

1. Remove Power and Antenna.
2. Remove screws and open the case.
3. Locate Synthesizer chip labeled MB8719
4. Cut wires connecting channel switch and pins 10-16.
5. Solder an on/off switch to each pin (pin 10-16)
6. reassemble radio.

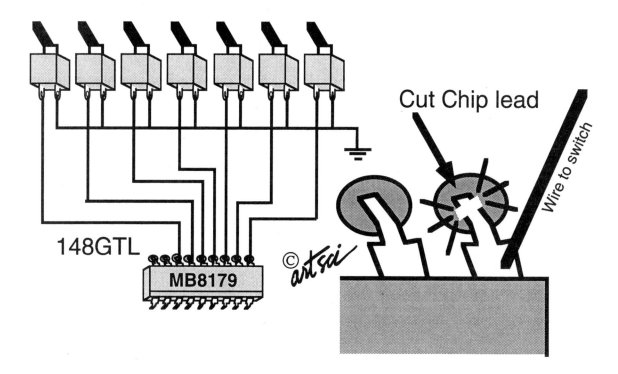

Frequency		10	11	12	13	14	15	16
26.815	=	1	0	0	0	0	0	0
26.825	=	1	0	0	0	0	0	1
26.835	=	1	0	0	0	0	1	0
26.845	=	1	0	0	0	0	1	1
26.855	=	1	0	0	0	1	0	0
26.865	=	1	0	0	0	1	0	1
26.875	=	1	0	0	0	1	1	0
26.885	=	1	0	0	0	1	1	1
26.895	=	1	0	0	1	0	0	0
26.905	=	1	0	0	1	0	0	1
26.915	=	1	0	0	1	0	1	0
26.925	=	1	0	0	1	0	1	1
26.935	=	1	0	0	1	1	0	0
26.945	=	1	0	0	1	1	0	1
26.955	=	1	0	0	1	1	1	0
26.965	=	1	0	0	1	1	1	1
26.975	=	1	0	1	0	0	0	0
26.985	=	1	0	1	0	0	0	1
26.995	=	1	0	1	0	0	1	0
27.005	=	1	0	1	0	0	1	1
27.015	=	1	0	1	0	1	0	0
27.025	=	1	0	1	0	1	0	1
27.035	=	1	0	1	0	1	1	0
27.045	=	1	0	1	0	1	1	1
27.055	=	1	0	1	1	0	0	0
27.065	=	1	0	1	1	0	0	1
27.075	=	1	0	1	1	0	1	0
27.085	=	1	0	1	1	0	1	1
27.095	=	1	0	1	1	1	0	0
27.105	=	1	0	1	1	1	0	1
27.115	=	1	0	1	1	1	1	0
27.125	=	1	0	1	1	1	1	1
27.135	=	1	1	0	0	0	0	0
27.145	=	1	1	0	0	0	0	1
27.155	=	1	1	0	0	0	1	0
27.165	=	1	1	0	0	0	1	1
27.175	=	1	1	0	0	1	0	0
27.185	=	1	1	0	0	1	0	1
27.195	=	1	1	0	0	1	1	0
27.205	=	1	1	0	0	1	1	1
27.215	=	1	1	0	1	0	0	0
27.225	=	1	1	0	1	0	0	1
27.235	=	1	1	0	1	0	1	0
27.245	=	1	1	0	1	0	1	1
27.255	=	1	1	0	1	1	0	0
27.265	=	1	1	0	1	1	0	1
27.275	=	1	1	0	1	1	1	0
27.285	=	1	1	0	1	1	1	1
27.295	=	1	1	1	0	0	0	0
27.305	=	1	1	1	0	0	0	1
27.315	=	1	1	1	0	0	1	0
27.325	=	1	1	1	0	0	1	1
27.335	=	1	1	1	0	1	0	0
27.345	=	1	1	1	0	1	0	1
27.355	=	1	1	1	0	1	1	0
27.365	=	1	1	1	0	1	1	1
27.375	=	1	1	1	1	0	0	0
27.385	=	1	1	1	1	0	0	1
27.395	=	1	1	1	1	0	1	0
27.405	=	1	1	1	1	0	1	1
27.415	=	1	1	1	1	1	0	0
27.425	=	1	1	1	1	1	0	1
27.435	=	1	1	1	1	1	1	0
27.445	=	1	1	1	1	1	1	1

Frequency		10	11	12	13	14	15	16
27.455	=	0	0	0	0	0	0	0
27.465	=	0	0	0	0	0	0	1
27.475	=	0	0	0	0	0	1	0
27.485	=	0	0	0	0	0	1	1
27.495	=	0	0	0	0	1	0	0
27.505	=	0	0	0	0	1	0	1
27.515	=	0	0	0	0	1	1	0
27.525	=	0	0	0	0	1	1	1
27.535	=	0	0	0	1	0	0	0
27.545	=	0	0	0	1	0	0	1
27.555	=	0	0	0	1	0	1	0
27.565	=	0	0	0	1	0	1	1
27.575	=	0	0	0	1	1	0	0
27.585	=	0	0	0	1	1	0	1
27.595	=	0	0	0	1	1	1	0
27.605	=	0	0	0	1	1	1	1
27.615	=	0	0	1	0	0	0	0
27.625	=	0	0	1	0	0	0	1
27.635	=	0	0	1	0	0	1	0
27.645	=	0	0	1	0	0	1	1
27.655	=	0	0	1	0	1	0	0
27.665	=	0	0	1	0	1	0	1
27.675	=	0	0	1	0	1	1	0
27.685	=	0	0	1	0	1	1	1
27.695	=	0	0	1	1	0	0	0
27.705	=	0	0	1	1	0	0	1
27.715	=	0	0	1	1	0	1	0
27.725	=	0	0	1	1	0	1	1
27.735	=	0	0	1	1	1	0	0
27.745	=	0	0	1	1	1	0	1
27.755	=	0	0	1	1	1	1	0
27.765	=	0	0	1	1	1	1	1
27.775	=	0	1	0	0	0	0	0
27.785	=	0	1	0	0	0	0	1
27.795	=	0	1	0	0	0	1	0
27.805	=	0	1	0	0	0	1	1
27.815	=	0	1	0	0	1	0	0
27.825	=	0	1	0	0	1	0	1
27.835	=	0	1	0	0	1	1	0
27.845	=	0	1	0	0	1	1	1
27.855	=	0	1	0	1	0	0	0
27.865	=	0	1	0	1	0	0	1
27.875	=	0	1	0	1	0	1	0
27.885	=	0	1	0	1	0	1	1
27.895	=	0	1	0	1	1	0	0
27.905	=	0	1	0	1	1	0	1
27.915	=	0	1	0	1	1	1	0
27.925	=	0	1	0	1	1	1	1
27.935	=	0	1	1	0	0	0	0
27.945	=	0	1	1	0	0	0	1
27.955	=	0	1	1	0	0	1	0
27.965	=	0	1	1	0	0	1	1
27.975	=	0	1	1	0	1	0	0
27.985	=	0	1	1	0	1	0	1
27.995	=	0	1	1	0	1	1	0
28.005	=	0	1	1	0	1	1	1
28.015	=	0	1	1	1	0	0	0
28.025	=	0	1	1	1	0	0	1
28.035	=	0	1	1	1	0	1	0
28.045	=	0	1	1	1	0	1	1
28.055	=	0	1	1	1	1	0	0
28.065	=	0	1	1	1	1	0	1
28.075	=	0	1	1	1	1	1	0
28.085	=	0	1	1	1	1	1	1

TRUTH TABLE FOR MB8719 IC

COBRA CB's
REMOVE ALC CIRCUIT (Higher TX power)

1. Remove Power and Antenna.
2. Remove screws and open the case.
3. Locate the indicated part and remove it.
4. Reassemble radio.

MODEL	REMOVE THIS PART
18-LTD	R-87
19 PLUS	D-502
20 PLUS	VR-502
21 PLUS	D-20
21 GTL	TR-14
21 LTD	TR-14 OR D9
21 XLR	TR-20
25 GTL	TR-14
25 PLUS	D-20
27	X8
29 GTL	D-20
29 PLUS	R-79 OR D-20
31 PLUS	D-19
32 XLR	TR-18
33 PLUS	D-17
40 PLUS	VR-104
78 X	C-49
85	D-9
86 XLR	CD-9
87 GTL	VR-6
89 GTL	VR-6
89 XLR	VR-5
132 XLR	R-134 = AM R-130 = SSB'
135 XLR	R-134 = AM R-130 = SSB'
138 XLR	TR-23
139 XLR	R-132
140 GTL	TR-32
142 GTL	TR-32
148 DX	VR-14=AM & VR-12=SSB
148 GTL	TR-24
150 GTL	RV-14=AM & RV-4=SSB
1000 GTL	VR-6
2000 GTL	TR-24 & C-232
REMOTE CONTROL	D-401

REALISTIC CB's
REMOVE ALC CIRCUIT (Higher TX power)

1. Remove Power and Antenna.
2. Remove screws and open the case.
3. Locate the indicated part and remove it.
4. Reassemble radio.

MODEL	REMOVE THIS PART
TRC-417	Q-19
TRC-421	D-16
TRC-422	Q-11
TRC-432	Q-12
TRC-440	D-107
TRC-448	VR-5=AM & VR-204=SSB
TRC-449	VR-7=AM & CT-7=SSB
TRC-452	VR-207
TRC-454	VR-702
TRC-455	R-504
TRC-457	VR-7=AM & CT-7=SSB
TRC-461	VR-2
TRC-462	D-17
TRC-467	D-109
TRC-468	R-42
TRC-469	VR-5
TRC-473	D-17
TRC-410	Q-12
TRC-413	R-85
TRC-415	Q-7
TRC-427	C-78
TRC-428	R-90
TRC-433	Q-15
TRC-451	VR-5=AM & VR-6=SSB
TRC-453	R-146
21-1537	D-17

OTHER CB's
REMOVE ALC CIRCUIT (Higher TX power)

1. Remove Power and Antenna.
2. Remove screws and open the case.
3. Locate the indicated part and remove it.
4. Reassemble the radio.

COMPANY	MODEL	REMOVE THIS PART
ALARON	B4900	Q-201
AUDIOVOX	WINSOR	D-12
	100	D-12
	CB-930	RV-2
	CB-950	D-39
	CBH-990	RV-2
	CBR-9600	RV-105
BROWNING	BARON	R-134=AM & R-130=SSB
	BROWNIE	Q-13
	MARK III	R-38=AM & R-69=SSB
	SABRE	CD-11
	SST-2	CD-11
CLARICON	PRIVATEER	CR-107

OTHER CB's
CONTINUED

COMPANY	MODEL	REMOVE THIS PART
COLT	190	R-71
	222	C-228
	290	RV-2
	320 DX	RV-14=AM & RV-4=SSB
	320 FM	RV-14=AM & RV-4=SSB
	350	R-121
	390	RV-2
	480	RV-12=AM & RV-11=SSB
	485	RV-12=AM & RV-11=SSB
	800	RV-2
	1000	RV-12=AM & RV-11=SSB
	1200 DX	RV-14=AM & RV-4=SSB
CONVOY	CON-400	R-129
COURIER	BLAZER 40D	VR-9
	CARAVELLE 40D	R-504
	CENTURIAN 40	D-24
	CENTURION 40D	D-46
	CHIEF 23	X-8
	CONQUEROR	R-504
	GLADIATOR	D-46
	NIGHT RIDER 40	VR-301
	RANGLER 40	VR-301
	RENEGADE 40	VR-9
	ROGUE 40	VR-5
CRAIG	L101	R-226
	L-321	R-605=AM & R-20=SSB

OTHER CB's
CONTINUED

COMPANY	MODEL	REMOVE THIS PART
DAK	IX	Q-202
	X	Q-37 & Q-38
FANNON	12SF	R-76
	190 DF	VR-301
	182F	D-12
	184DF	D-12
	185DF	VR-301
	185PLL	VR-301
	SFT 400	D-10
FUZZBUSTER	2-50	Q-8
GE	3-5801A	VR-7
	3-5804A	VR-7
	3-5804D	RV-2
	3-5810B	RV-2
	3-5811B	RV-2
	3-5812A	RV-2
	3-5813A	RV-2
	3-5813B	RV-2
	3-5814A	C-98
	3-5814B	RV-2
	3-5818A	RV-2
	3-5819A	RV-2
	3-5821A	VR-10
	3-5821B	VR-10
	3-5869A	RV-2
	3-5871A	VR-11
	3-5871B	VR-11
	3-5875A	RV-9=AM & VR-201=SSB

OTHER CB's
CONTINUED

COMPANY	MODEL	REMOVE THIS PART
GEMTRONICS	GTX-44	RV-2
	GTX-55	RV-2
	GTX-66	RV-2
	GTX-77	RV-2
	3000-GTX	R-93
	4040	D-481
	5000-GTX	VR-4
HY-GAIN	672 B	RV-2
	674 B	VR-7
	2679 I	RV-2
	2680 II	RV-2
	2681 II	RV-2
	2682 II	RV-2
	2683 III	RV-2
	2701 I	RV-2
	2702 II	RV-2
	2703 III	RV-2
	2795	RV-14=AM & RV-4=SSB
	2795 DX	RV-14=AM & RV-4=SSB
	V SSB	VR-7
JC PENNY	981-6221	D-501
	981-6237	D-7
	681-6241	Q-405
	6218	RV-2

OTHER CB's
CONTINUED

COMPANY	MODEL	REMOVE THIS PART
JOHNSON	4120	CR-12
	4125	CR-12
	4135	CR-12
	4140	R-37
	4145	R-37
	4230	R-37
KRACO	KCB-4000	VR-4
	KCB-4010	RV-2
	KCB-4020	RV-2
	KCB-4030	RV-2
	KCB-4045	RV-2
LAFAYETTE	HB-650	RV-102
	HB-750	RV-102
	HB-870	RV-14=AM & RV-4=SSB
	HB-940	RV-2
	SSB-100	RV-7=AM & RV-8=SSB
	SSB-140	RV-12=AM & RV-11=SSB
	TELSTAT 1140	RV-2
	TELSTAT 1240	VR-305

OTHER CB's
CONTINUED

COMPANY	MODEL	REMOVE THIS PART
MIDLAND	76-858	RV-2
	76-860	R-218
	76-863	RV-2
	77-101B	RV-201
	77-101C	RV-201
	77-116	RV-2
	77-821	RV-2
	77-824	RV-201
	77-825	D-3
	77-830	RV-2
	77-838	RV-2
	77-849	RV-2
	77-856	VR-5
	77-857	RV-2
	77-861	D-2
	77-866	TR-8
	77-867	D-14
	77-874	X-11
	77-882	Q-15
	77-883	X-11
	77-888	RV-2
	77-889	RV-2
	77-963	RV-2
	79-892	RV-12=AM & RV-11=SSB
	79-893	RT-601=AM & RV-7=SSB
MOPAR	4094177	RV-2
	4094178	RV-2

OTHER CB's
CONTINUED

COMPANY	MODEL	REMOVE THIS PART
PACE	CB-145	CV-20
	CB-166	R-207
	1000-MS	CR-508
	2300	X-9
	CB-8008	R-218
	CB-8010	R-220
	CB-8015	R-220
	CB-8041	R-302
	CB-8046	R-302
	CB-8117	R-220
	CB-8117	R-220
PALOMAR	49	R-208
	SSB-500	RV-12=AM & RV-2=SSB
	4100	RV-2
PANASONIC	RJ-3150	R-117
	RJ-3250	R-70
PEARCE	JAGUAR	FVR-3
SIMPSON	LION	RV-2
	SUPER LYNX	D-12
	TIGER	RV-2

OTHER CB's
CONTINUED

COMPANY	MODEL	REMOVE THIS PART
PRESIDENT	ADAMS (OLD)	VR-7=AM & CT-7=SSB
	ADAMS (NEW)	TR-24
	AR-7	R-54
	AX-43	Q-12
	DWIGHT D	VR-6
	GRANT (OLD)	VR-7=AM & CT-7=SSB
	GRANT (NEW)	R-128=AM & VR-11=SSB
	HONEST ABE	VR-5
	JOHN Q	RT-4
	MADISON (OLD)	VR-7=AB & CT-7=SSB
	MADISON (NEW)	R-128
	MCKINLEY	R-120
	OLD HICKORY	VR-5
	TEDDY R	VR-5
	THOMAS J	VR-4
	WASHINGTON (OLD)	VR-7=AM & CT-7=SSB
	WASHINGTON (NEW)	TR-32
	ZACHARY T	VR-6
RAIDER	404-R	D-52
RANGER	AR-3300	VR-17=AM & VR-15=SSB
	AR-3500	VR-17=AM & VR-15=SSB
RCA	14T260	RV-2
	14T270	RV-2
	14T301	RV-2
	14T302	D-301
	14T303	RV-2
	14T304	RV-2
	14T305	RV-2

OTHER CB's
CONTINUED

COMPANY	MODEL	REMOVE THIS PART
RCI	2900	VR-14=AM & VR-12=SSB
	2950	VR-14=AM & VR-12=SSB
REGENCY	CR-186	D-9
ROBYN	AM-500D	VR-5
	DG-130D	VR-6
	GT-410	VR-13
	LB-120	VR-6
	SX-401	RV-7
	SX-402D	VR-13
	T240D	VR-4
	WV-110	VR-6
	007-140	VR-6
	123-C	D-11
	510-D	VR-7=AM & CT-7=SSB

OTHER CB's
CONTINUED

COMPANY	MODEL	REMOVE THIS PART
ROYCE	1-602	D-6
	1-603	Q-205
	1-606	D-17
	1-607	VR-201
	1-609	Q-205
	1-610	D-202
	1-619	D-301
	1-620	D-301
	1-621	VR-3
	1-625	VR-1602
	1-630	C-79 & D-42 & D-44
	1-639	Q-16
	1-641	VR-7
	1-648	C-82 & C-35 & C-96
	1-653D	D-301
	1-655	D-301
	1-658	D-301
	1-662	D-301
	1-673	D-301
	1-675	D-301
	1-680	D-301
	1-682	D-301
SANYO	TA-2000	D-505
	TA-4000	VR-6

OTHER CB's
CONTINUED

COMPANY	MODEL	REMOVE THIS PART
SBE	ASPEN-41	VR-203
	CONSOLE II	VR-7=AM & VR-1=SSB
	CONSOLE V	VR-803=AM & VR-302=SSB
	CORTEX	VR-203
	FORMULA D	VR-9
	KEYCOM 54	RV-1
	LCB-8	VR-6
	LCMS-5	VR-6
	MALIBU 44	R-226
	TAHOE 49	R-129
	TOUCH COM 174	VR-4
	TRINIDAD 45	R-226
SEARS	370.380507	R-218
	934.36710501	D-6
	934.380607	D-7
	934.380627	R-42
	934.380807	D-7
	934.380817	D-501
	934.381107	D-501
	934.381207	D-502
SILTRONICS	APACHE	D-14
	MOHAWK	D-14
SHARP	CB-750	R-112
	CB-2260	R-112

OTHER CB's
CONTINUED

COMPANY	MODEL	REMOVE THIS PART
SUPERSTAR	1 2 0	D-11
	360 FM	VR-14=AM & VR-12=SSB
	3600	VR-14=AM & VR-12=SSB
TEABERRY	RACER T	VR-6
	STALKER I	VR-13=AM & VR-12=SSB
	STALKER II	VR-13=AM & VR-12=SSB
	STALKER V	VR-4
	STALKER IX	R-102
	T BEAR	VR-5
	T CHARLIE	VR-7
	T COMMAND	VR-5
	TITAN T	D-14
	T CONTROL	VR-505
TENNA PHASE	CB-22	R-46
	CB-26	D-22
TRAM	D-12	R-61
	D-42	CD-11
	D-60	R-98=AM & R-112 SSB
	D-201A	VR-77
	D-300	TR-23
TRUETONE	CYJ4862A-87	RV-2
	8334	Q-15

OTHER CB's
CONTINUED

COMPANY	**MODEL**	**REMOVE THIS PART**
UNIDEN	2510	VR-112=AM & VR-104=SSB
	2600	VR-112=AM & VR-104=SSB
	PC-3	TR-14
	PRO-640	RV-5=AM & VR-6=SSB
	PC-122	Q-29 (near PL connector)
UTAC	TRX-400	D-11
VECTOR	770	FVR-3
	790	FVR-3
WARDS	GEN-730A	VR-206
	GEN-775A	VR-206
	GEN-828A	VR-206
WHISTLER	700	Q-205
	900	Q-305
XTAL	CB-7	D-18
	CB-11	D-14
	SSB-10	D-2
ZEXON	49	Q-201

Radio / Tech Modifications

APPENDIX

COAX TYPE	VEL %	dB ATTENUATION PER 100 FEET.				LENGTH IN FEET FOR 1 WAVELENGHT		
		100 MHz	200 Mhz	400 MHz	1000 MHz	146 MHz	222 MHz	445 MHz
9913 (100% shield)	89	1.4	1.8	2.6	4.5	6.00	3.94	1.97
RG-8U FOAM (8214)	80	1.8	2.7	4.2	7.0	5.39	3.55	1.77
RG-213 (NON-CONTAM.)	66	2.2	3.2	4.7	8.5	4.45	2.93	1.46
RG-8X (MINI-FOAM)	78	3.7	5.4	8.0	13.5	5.26	3.46	1.72
9311 (100% SHIELD 58U)	78	4.5	6.3	9.0	14.5	5.26	3.46	1.72
RG-58U (SOLID CENTER)	66	4.5	6.7	10.0	17.0	4.45	2.93	1.46
RG-58A/U (STRANDED CTR)	66	4.9	7.5	11.5	21.5	4.45	2.93	1.46

(c) 1989 N6MOS

COAX LENGTHS SHOULD BE MULTIPLE HALF WAVELENGHTS.

$$\frac{984}{\text{FREQ. IN MHZ}} \times \text{VEL \%} = \text{ONE WAVE LENGHT IN FEET.}$$

db - % loss chart

db Loss	Power Loss	db Loss	Power Loss	db Loss	Power Loss	db Loss	Power Loss
		2.0	37 %	4.0	60 %	6.0	75 %
0.2	4 %	2.2	39 %	4.2	62 %	7.0	80 %
0.4	8 %	2.4	42 %	4.4	63 %	8.0	84 %
0.6	13 %	2.6	45 %	4.6	65 %	9.0	87 %
0.8	17 %	2.8	47 %	4.8	67 %	10.0	90 %
1.0	21 %	3.0	50 %	5.0	68 %	20.0	99 %
1.2	24 %	3.2	52 %	5.2	70 %	30.0	100 %
1.4	27 %	3.4	54 %	5.4	71 %	40.0	100 %
1.6	30 %	3.6	56 %	5.6	73 %		
1.8	33 %	3.8	58 %	5.8	74 %		

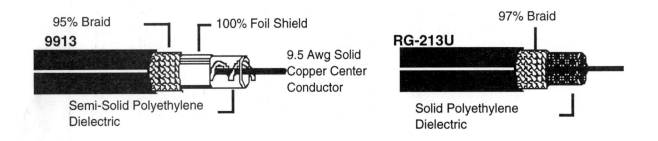

95% Braid
9913
100% Foil Shield
9.5 Awg Solid Copper Center Conductor
Semi-Solid Polyethylene Dielectric

97% Braid
RG-213U
Solid Polyethylene Dielectric

APPENDIX A

RESISTOR COLOR CODE

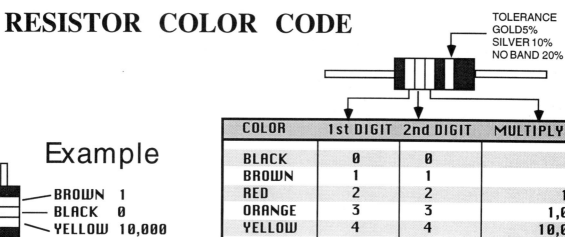

TOLERANCE
GOLD 5%
SILVER 10%
NO BAND 20%

Example

BROWN 1
BLACK 0
YELLOW 10,000
SILVER 10%

10 X 10,000 = 100,000
(100K) OHMS

COLOR	1st DIGIT	2nd DIGIT	MULTIPLY BY
BLACK	0	0	1
BROWN	1	1	10
RED	2	2	100
ORANGE	3	3	1,000
YELLOW	4	4	10,000
GREEN	5	5	100,000
BLUE	6	6	1,000,000
VIOLET	7	7	10,000,000
GRAY	8	8	100,000,000
WHITE	9	9	1,000,000,000
GOLD			.1
SILVER			.01

CAPACITORS

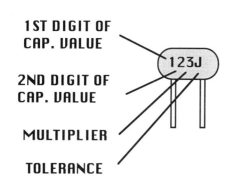

1ST DIGIT OF CAP. VALUE

2ND DIGIT OF CAP. VALUE

MULTIPLIER

TOLERANCE

123J

MULTIPLIER		TOLERANCE	
	MULTIPLY BY		10pF or less	over 10pF
0	1	B	0.1pF	
1	10	C	0.25pF	
2	100	D	0.5pF	
3	1,000	F	1.0pf	1%
4	10,000	G	2.0pf	2%
5	100,000	H		3%
		J		5%
8	.01	K		10%
9	0.1	M		20%

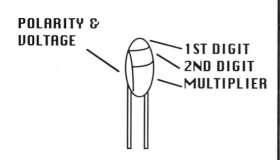

POLARITY & VOLTAGE

1ST DIGIT
2ND DIGIT
MULTIPLIER

COLOR	DIGIT	MULTIPLIER	VOLTAGE
BLACK	0	NONE	4
BROWN	1	10	6
RED	2	100	10
ORANGE	3	1,000	15
YELLOW	4	10,000	20
GREEN	5	100,000	25
BLUE	6	1,000,000	35
VIOLET	7	10,000,000	50
GRAY	8		
WHITE	9		

APPENDIX B

PL ENCODER HOOK-UP

PL Encoder Connections

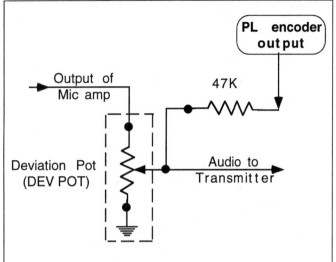

Attach a 47K ohm resistor to the output of the deviation pot. Attach the other end of resistor to the output of the PL encoder.

PL TONE CHART

PL TONE	FREQ. CODE	ICOM	YAESU	TS-32 SWITCH 1 2 3 4 5
67.0	-XZ	1	1	1 1 1 1 1
71.9	-XA	2	2	0 1 1 1 1
74.4	-WA	3	36	1 0 1 1 1
77.0	-XB	4	3	0 0 1 1 1
79.7	-SP	5	38	1 1 0 1 1
82.5	-YZ	6	4	0 1 0 1 1
85.4	-YA	7	40	1 0 0 1 1
88.5	-YB	8	5	0 0 0 1 1
91.5	-ZZ	9	42	1 1 1 0 1
94.8	-ZA	10	6	0 1 1 0 1
97.4	-ZB	11		1 0 1 0 1
100.0	-1Z	12	7	0 0 1 0 1
103.5	-1A	13	8	1 1 0 0 1
107.2	-1B	14	9	0 1 0 0 1
110.9	-2Z	15	10	1 0 0 0 1
114.8	-2A	16	11	0 0 0 0 1
118.8	-2B	17	12	1 1 1 1 0
123.0	-3Z	18	13	0 1 1 1 0
127.3	-3A	19	14	1 0 1 1 0
131.8	-3B	20	15	0 0 1 1 0
136.5	-4Z	21	16	1 1 0 1 0
141.3	-4A	22	17	0 1 0 1 0
146.3	-4B	23	18	1 0 0 1 0
151.4	-5Z	24	19	0 0 0 1 0
156.7	-5A	25	20	1 1 1 0 0
162.2	-5B	26	21	0 1 1 0 0
167.9	-6Z	27	22	1 0 1 0 0
173.8	-6A	28	23	0 0 1 0 0
179.9	-6B	29	24	1 1 0 0 0
186.2	-7Z	30	25	0 1 0 0 0
192.8	-7A	31	26	1 0 0 0 0
203.5	-M1	32	27	0 0 0 0 0
210.7		33		

1 = on / 2 = off
example above
01001=107.2

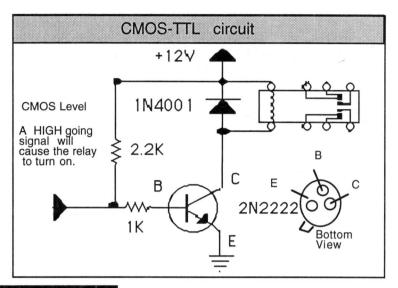

CMOS-TTL circuit

+12V

1N4001

CMOS Level

A HIGH going signal will cause the relay to turn on.

2.2K

B

C

1K

B

2N2222

E

C

E

Bottom View

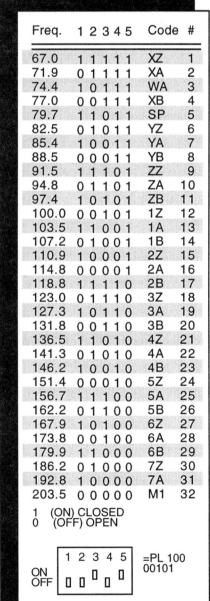

Freq.	1 2 3 4 5	Code	#
67.0	1 1 1 1 1	XZ	1
71.9	0 1 1 1 1	XA	2
74.4	1 0 1 1 1	WA	3
77.0	0 0 1 1 1	XB	4
79.7	1 1 0 1 1	SP	5
82.5	0 1 0 1 1	YZ	6
85.4	1 0 0 1 1	YA	7
88.5	0 0 0 1 1	YB	8
91.5	1 1 1 0 1	ZZ	9
94.8	0 1 1 0 1	ZA	10
97.4	1 0 1 0 1	ZB	11
100.0	0 0 1 0 1	1Z	12
103.5	1 1 0 0 1	1A	13
107.2	0 1 0 0 1	1B	14
110.9	1 0 0 0 1	2Z	15
114.8	0 0 0 0 1	2A	16
118.8	1 1 1 1 0	2B	17
123.0	0 1 1 1 0	3Z	18
127.3	1 0 1 1 0	3A	19
131.8	0 0 1 1 0	3B	20
136.5	1 1 0 1 0	4Z	21
141.3	0 1 0 1 0	4A	22
146.2	1 0 0 1 0	4B	23
151.4	0 0 0 1 0	5Z	24
156.7	1 1 1 0 0	5A	25
162.2	0 1 1 0 0	5B	26
167.9	1 0 1 0 0	6Z	27
173.8	0 0 1 0 0	6A	28
179.9	1 1 0 0 0	6B	29
186.2	0 1 0 0 0	7Z	30
192.8	1 0 0 0 0	7A	31
203.5	0 0 0 0 0	M1	32

1 (ON) CLOSED
0 (OFF) OPEN

1 2 3 4 5   =PL 100
ON   00101
OFF

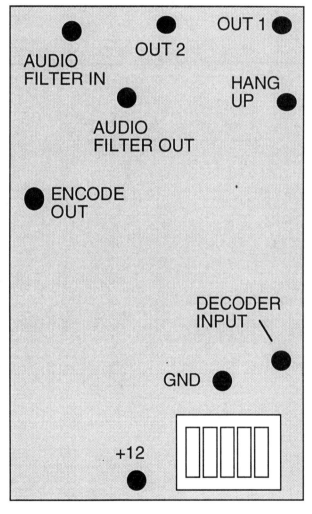

OUT 1

OUT 2

AUDIO FILTER IN

HANG UP

AUDIO FILTER OUT

ENCODE OUT

DECODER INPUT

GND

+12

TS-32 LAYOUT

APPENDIX D

TS-32 HOOKUP
PL Decoder

WHEN THE SELECTED PL TONE IS RECEIVED, THE RELAY WILL CLOSE
AND AUDIO WILL BE PASSED TO THE SPEAKER.

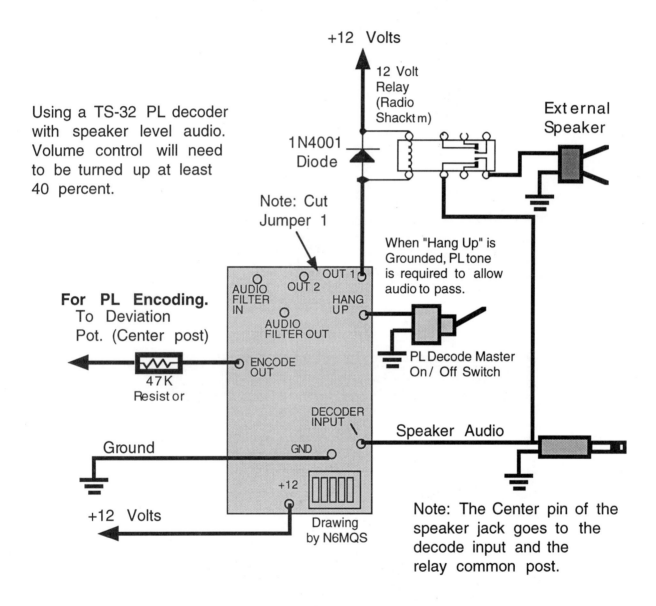

Using a TS-32 PL decoder
with speaker level audio.
Volume control will need
to be turned up at least
40 percent.

+12 Volts

12 Volt
Relay
(Radio
Shack tm)

External
Speaker

1N4001
Diode

Note: Cut
Jumper 1

When "Hang Up" is
Grounded, PL tone
is required to allow
audio to pass.

For PL Encoding.
To Deviation
Pot. (Center post)

AUDIO
FILTER
IN

OUT 1

OUT 2

HANG
UP

AUDIO
FILTER OUT

ENCODE
OUT

47K
Resistor

PL Decode Master
On / Off Switch

DECODER
INPUT

Speaker Audio

Ground

GND

+12

+12 Volts

Drawing
by N6MQS

Note: The Center pin of the
speaker jack goes to the
decode input and the
relay common post.

APPENDIX   E

PL DECODER HOOK-UP

PL Decoder Connections

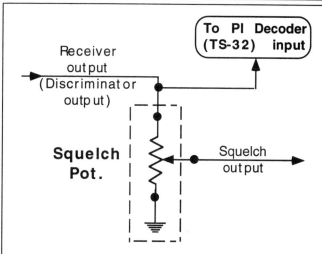

Attach a wire to the discriminator output. Attach the other end to the decoder input. The discriminator output is often connected to the squelch pot. See audio connections below for audio control.

PL Decoder/Audio Connections

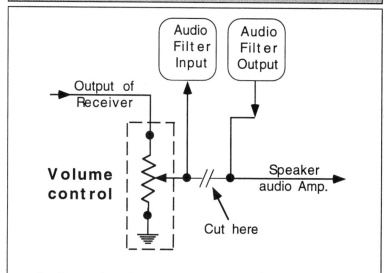

Audio muting is controlled by the TS-32 Board. When a PL is present on the signal, audio will pass.

##	FREQ.	DESCRIPTION	HAN ##	FREQ.	DESCRIPTION
1			51		
2			52		
3			53		
4			54		
5			55		
6			56		
7			57		
8			58		
9			59		
10			60		
11			61		
12			62		
13			63		
14			64		
15			65		
16			66		
17			67		
18			68		
19			69		
20			70		
21			71		
22			72		
23			73		
24			74		
25			75		
26			76		
27			77		
28			78		
29			79		
30			80		
31			81		
32			82		
33			83		
34			84		
35			85		
36			86		
37			87		
38			88		
39			89		
40			90		
41			91		
42			92		
43			93		
44			94		
45			95		
46			96		
47			97		
48			98		
49			99		
50			100		

APPENDIX G

Performance Report

Radio _____   Date _____

Owner : Name _____
            Address _____
            City _____ St. _____ Zip _____
            Phone (____) _____ - _____

Description	Before	After
Power out (Low)	_____ Watts	_____ Watts
Power out (High)	_____ Watts	_____ Watts
Frequency Error (Simplex)	_____ Hz	_____ Hz
Frequency Error (Offset)	_____ Hz	_____ Hz
Receive Sensitivity (Mid-band)	_____ uv	_____ uv
Receive Sensitivity (____MHz)	_____ uv	_____ uv
Receive Sensitivity (____MHz)	_____ uv	_____ uv
PL Deviation	_____ Hz	_____ Hz
DTMF Deviation	_____ KHz	_____ KHz
Audio Deviation	_____ KHz	_____ KHz
Lowest usable Freq @ .5 Pwr	_____ MHz	_____ MHz
Highest usable Freq @ .5 Pwr	_____ MHz	_____ MHz

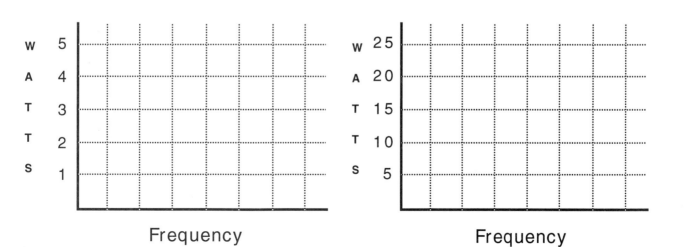

Frequency   Frequency

Performance Report

Radio _____   Date _____

Owner : Name _____
        Address _____
        City _____ St. ____ Zip _____
        Phone (____) _____ - _____

Description	Before	After
Power out (Low)	_____ Watts	_____ Watts
Power out (High)	_____ Watts	_____ Watts
Frequency Error (Simplex)	_____ Hz	_____ Hz
Frequency Error (Offset)	_____ Hz	_____ Hz
Receive Sensitivity (Mid-band)	_____ uv	_____ uv
Receive Sensitivity (____MHz)	_____ uv	_____ uv
Receive Sensitivity (____MHz)	_____ uv	_____ uv
PL Deviation	_____ Hz	_____ Hz
DTMF Deviation	_____ KHz	_____ KHz
Audio Deviation	_____ KHz	_____ KHz
Lowest usable Freq @ .5 Pwr	_____ MHz	_____ MHz
Highest usable Freq @ .5 Pwr	_____ MHz	_____ MHz

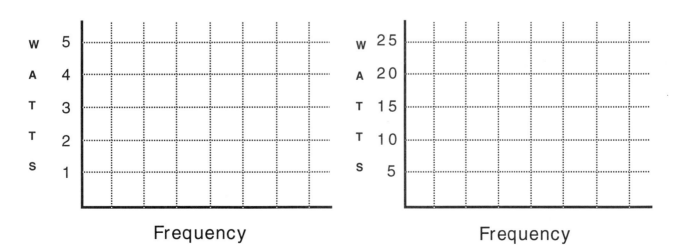

Frequency   Frequency

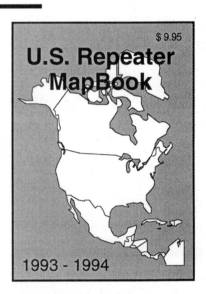
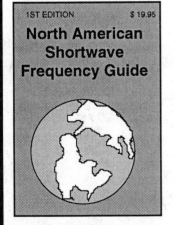

RIDING THE AIRWAVES WITH ALPHA & ZULU

Study for the Novice and No-Code Technician license tests with the newest comic book characters the Phoneticos.
112 comic strips review all the questions and answers.
288 pages, 8 1/2 X 11" format

U.S. REPEATER MAPBOOK #2

A repeater guide that shows where in each state principal open amateur repeaters are located. The Maps also show the important highways in each state. Tables showing the popular repeater in the states major cities are also presented. 2 meter, 200, 440 MHz and 1.2 GHz repeaters are shown. 144 pages, 6 x 9" format

FEDERAL ASSIGNMENTS Vol 3

The Frequency assignment master file.
The complete listing of all U.S. government used frequencies listed by agency and in frequency order.
Frequencies for Departments of: Agriculture, Air Force, Army, Commerce, Defence, Energy, Health and Human Services, Housing and Urban Developement, Interior, Justice, Labor, Navy, State, Treasury, Transportation and 29 Independent agencies & Commissions.
Over 350 pages, 8 1/2 X 11" format

AMATEUR HAMBOOK

Equipment & Log Sheets, Charts, Tables showing: worldwide callsigns, world times, shortwave listening frequencies, coax losses, CTCSS details, conversions, construction plans, emergency information, etc.
This book contains all the useful information a amateur radio operator needs to reference.
133 pages, 6 X 9" format.

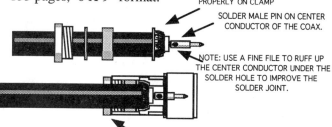

TRIM BRAID TO FIT PROPERLY ON CLAMP

SOLDER MALE PIN ON CENTER CONDUCTOR OF THE COAX.

NOTE: USE A FINE FILE TO RUFF UP THE CENTER CONDUCTOR UNDER THE SOLDER HOLE TO IMPROVE THE SOLDER JOINT.

PUSH PLUG BODY ON COAX AND TIGHTEN THE NUT.

QUICK-N-EASY SHORTWAVE LISTENING

What kind of radio to Buy? Whats a good antenna, What is there to listen to? This book contains pictures of radios, antenna construction and frequency lists. 6 X 9" format

P.O. Box 1428
Burbank, CA 91507
(818) 843-4080
FAX: (818) 846-2298

RADIO/TECH MODIFICATIONS # 5A

Modifications and alignment controls for ICOM & KENWOOD amateur radios, and UNIDEN, RADIO SHACK & REGENCY Scanners. Over 200 pages, 8 1/2" X 11"

RADIO/TECH MODIFICATIONS # 5B

Modifications and alignment controls for ALINCO, YAESU, STANDARD, AZDEN radios and 10 meter & CB radios.

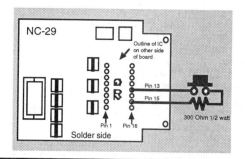

HAM RADIO RESOURCE GUIDE

For Southern California only. A booklet of all the information an amateur radio operator needs.
Listings of clubs, testing centers, sario stores and surplus dealers. Maps of repeaters, store and swap meet locations. Listing of Packet repeaters, phone BBS and node lists. 64 pages 8 1/2" X 5 1/2"

NORTH AMERICAN SHORTWAVE FREQUENCY GUIDE

Accurate and complete listing of all English and Spanish broadcasts on the 0 - 30 MHz shortwave bands. Listing are presented in frequency order.
Over 200 pages, 8 1/2" X 11"

LOST USER MANUALS

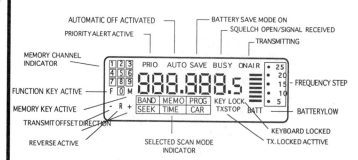

Lost the manual for your HT or Mobile rig? Did you purchase a used radio and it did not come with a manual? Do you have the manual but still can work the radio quickly?
"LOST USERS MANUALS" contains operating instructions for all the popular amateur radios and scanners. ICOM, Yaesu, Kenwood, Alinco, Standard, Uniden and other manufactors radios. Each radio is given 2 to 5 pages of drawing, charts and programming instructions. Over 140 Pages, 8 1/2 X 11" format.

ORDER FORM

	TITLE	DESCRIPTION	PRICE	QTY	EXTENSION
Radio Reference	Radio/Tech Modifications VOL 6A	Over 200 pages of mods for ICOM, KENWOOD Radios & all models of Scanners	19.95		
	Radio/Tech Modifications VOL 6B	Over 200 pages of mods for ALINCO, YAESU, STANDARD and all models of CB equipment.	19.95		
	Lost Users Manuals	Operating Instructions for all popular amateur Mobiles & Ht's.	19.95		
Amateur Reference	U.S. Repeater Mapbook	VHF & UHF Repeater guide for the USA with State Maps showing popular repeaters.	9.95		
	Amateur HamBook #2	Construction plans, coax, antenna, connector, SWR charts. A must have.	14.95		
	Ham Radio Resource Guide	For Southern California only. Testing, Club, Repeater, maps & more	6.00		
Scanner/Freq. Reference	Federal Assignments Volume #3	Scanner Frequency guide for all Federal Government Agencies. Over 300 pages	24.95		
	Police & Fire Communications Handbook	For Southern & Central California Scanner Listners. Best available freq. List !!	16.95		
Shortwave Freq. & Ref	Quick-N-Easy Shortwave Listening	Beginning guide to getting started with shortwave listening. Antennas Accessories, Receivers & more.	9.95		
	North American Shortwave Directory	Complete Listing of all activity on the HF band 0-30MHz.	19.95		
License Study Guide	Riding the airwaves with Alpha & Zulu	Novice & No-code license test book using cartoon strips to teach. for ages 8 - 80 !!!	14.95		

MAIL ORDER FORM TO:

ARTSCI INC.
P.O. BOX 1428
BURBANK, CA 91507
(818) 843-4080
FAX: (818) 846-2298

Shipping charge outside the U.S. is $10.00 or more

	EXTENSION
SUBTOTAL	
SALES TAX 8.25% CA	
SHIPPING	$ 4.00
ORDER TOTAL	

SHIP TO:

NAME

ADDRESS

CITY   ST   ZIP

PHONE (   )

BILLING INFORMATION   Give us your phone number in case we have a problem processing your order.

☐ CHECK ENCLOSED

☐ VISA / MasterCharge / DISCOVER / AMERICAN EXPRESS

artsci

CARD #

EXP DATE